# Statistics

# Workbook

**Created by Benjamin Etgen**

**and David Straayer of Tacoma Community College**

*Required for Probability and Statistics (Stat 300)*

*With Professor Etgen*

# Table of Contents

# Chapter 1
## Statistical Basics

The mathematical path from data to knowledge.

**Statistics** is the study of how to collect, organize, analyze, and interpret data.

Two branches:

descriptive statistics ( _____ & _____ data)

inferential statistics ( _____ & _____ data)

### 1.1: Some key terms

**Individual** – a person or object that you are interested in finding out information about.

**Variable** (also known as a **random variable**) – the measurement or observation of the individual.

**Population** – set of all the entire group of individuals about which we are interested.

**Sample** – a subset from the population. It looks just like the population, but contains less data.

Parameters and Statistics

**Parameter** – a number calculated from the _____. Usually denoted with a Greek letter. This number is a fixed, unknown number that you want to find.

**Statistic** – a number calculated from the _____. Usually denoted with letters from the Latin alphabet, though sometimes there is a Latin letter with a ^ (caret called a "hat") above it. Sometimes we just put a bar over the letter. Since you can find samples, statistics are readily known. Statistics change depending on the sample taken. Statistics are used to estimate the parameter value.

# Types of variables

**Qualitative or categorical variable** – is a word or name that $\bigcirc$ a quality of the individual.

**Quantitative or numerical variable** – is a number, something that can be counted or $\bigcirc$ from the individual. If it's a number, it's probably quantitative – unless it makes no sense to do arithmetic.

# Measurement Scales

**Nominal** – data is just a $\bigcirc$ or $\bigcirc$. There is no order to any data and since there are no numbers, you cannot do any arithmetic on this level of data. Examples of this are gender, car name, ethnicity, and race.

**Ordinal** – data that is nominal, but you can now put the data in $\bigcirc$, since one value is more or less than another value. You cannot do arithmetic on this data, but you can now put data values in order. Examples of this are grades (A, B, C, D, F), place value in a race (1st, 2nd, 3rd), and size of a drink (small, medium, large).

**Interval** – data that is ordinal, but you can now $\bigcirc$ one value from another and that subtraction makes sense. You can do arithmetic on this data, but only addition and subtraction. Examples of this are temperature and time on a clock.

**Ratio** – data that is interval, but you can now $\bigcirc$ one value by another and that ratio makes sense. You can now do all arithmetic on this data. Examples of this are height, weight, distance, and time.

# Families of Numbers

Counting numbers: Integers, Whole numbers, Natural Numbers

Measuring numbers: Fractions, decimals, scientific notation, …

**Section 1.2: Sampling Methods**

**census** – not really a sample (Try to measure ⬭)

**simple random sample** – every different possible sample of size *n* has the ⬭ chance of being selected.

**stratified sample** – ⬭ into strata, randomly select in each.

**systematic sample** – ⬭ $n^{th}$, for example.

**cluster sample** – divide into ⬭, randomly ⬭ clusters, sample some or all in ⬭ clusters.

**convenience sample** – not statistically valid.

The problem of Bias

Bias means that a sample does not represent the population from which it was drawn. Hence, conclusions from the sample may not apply to the population as a whole. Biases can be anticipated or unanticipated. *Randomization* is the key to avoid bias.

**Section 1.3: Experimental Design**

**Observational Study** – when the investigator collects data merely by ⬭ or ⬭⬭, without changing anything.

**Experiment** – when the investigator ⬭ a variable or imposes a ⬭ to determine its effect.

# Experimental Design Guidelines for Planning

1) Identify individuals of interest.

2) Specify variable (or variables).

3) Specify population.

4) Specify method for measuring or observing.

5) Determine sampling method.

6) Collect data.

7) Use inference to come knowledge.

8) Refine (problems, recommendations)

## Observation vs Experiment

**Observational Study** – when investigators collect data merely by watching or asking questions. They don't change anything. Very ⬯ to establish cause-and-effect.

**Experiment** – when the investigator changes a variable or imposes a treatment to determine its effect. ⬯ to establish cause-and-effect.

## Experimental Options

**Randomized two-treatment experiment** – *Control* is key concept. *Placebo* is often used for control.

**Randomized Block Design** – a block is a group of subjects that are ⬯, but the blocks differ from each other. Then ⬯ assign treatments to subjects inside each block

**Rigorously Controlled Design** – carefully assign subjects to different treatment groups, so that those given each treatment are similar in ways that are ⬯ to the experiment

**Matched Pairs Design** – the treatments are given to (＿＿＿＿＿＿) groups that can be matched up with each other in some ways.

Other Aspects of Experimentation

**Replication** Nobody is going to make major changes based on a (＿＿＿＿＿＿) study.

**Blinding**

Single blinded: subjects does not know if (＿＿＿＿＿) or (＿＿＿＿＿)

Double blinded: Those (＿＿＿＿＿) doing the measuring do not know either.

Experiments and Time

**Cross-sectional study** – data observed, measured, or collected at (＿＿＿＿＿) (＿＿＿＿＿) in time.

**Retrospective (or case-control) study** – data collected from the (＿＿＿＿＿) using (＿＿＿＿), (＿＿＿＿＿), and other similar artifacts.

**Prospective (or longitudinal or cohort) study** – data collected in the (＿＿＿＿＿) from groups sharing common factors.

**Section 1.4: How Not to Do Statistics**

**lurking or confounding variables** – when you cannot rule out the possibility that the observed effect is due to some other (＿＿＿＿＿) rather than the factor being studied.

**Overgeneralization** – where you do a study on one group and then try to say that it will happen on (＿＿＿＿＿) groups.

## Cause and Effect

We can not say that one variable causes the other just because the variables are related or $\bigcirc$.

It is **very** difficult to establish cause from $\bigcirc$ alone.

## Other Issues

**Sampling error** – This is the difference between the $\bigcirc$ results and the true $\bigcirc$ results.

**Nonsampling error** – This is where the sample is collected $\bigcirc$ either through a biased sample or through error in measurements. Care should be taken to avoid this error.

### Significance – two meanings

Statistical Significance – The results are probably $\bigcirc$ due to the chance of sampling

Practical or Clinical Significance – Does the effect really $\bigcirc$?

## Bias in Surveys

The wording of the questions can cause **hidden bias** – where the questions are asked in a way that makes a person respond a certain way

**Non-response** – where you send out a survey but $\bigcirc$ everyone returns the survey

**Voluntary response** – where people are asked to respond via phone, email or online. The problem with these is that only people who really $\bigcirc$ about the topic are likely to call or email.

# Chapter 2

## Graphical Descriptions of Data

### Section 2.1: Categorical Data

Frequency tables

Bar charts

Pie Charts

Pareto Charts

Multiple Bar Graphs

Frequency tables

Typically, the different categories are different rows.

One column might be frequency counts

Sometimes we replace or add a relative frequency column

| Category | Frequency | | Category | Frequency | Relative Frequency |
|---|---|---|---|---|---|
| Ford | 5 | | Ford | 5 | 0.10 |
| Chevy | 12 | | Chevy | 12 | 0.24 |
| Honda | 6 | | Honda | 6 | 0.12 |
| Toyota | 12 | | Toyota | 12 | 0.24 |
| Nissan | 10 | | Nissan | 10 | 0.20 |
| Other | 5 | | Other | 5 | 0.10 |
| Total | 50 | | Total | 50 | 1.00 |

## Bar Graphs or Charts

Put the frequency on the vertical axis and the category on the horizontal axis.

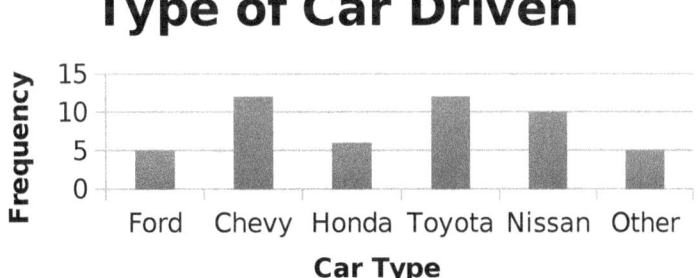

Key Features of Bar graphs

Equal 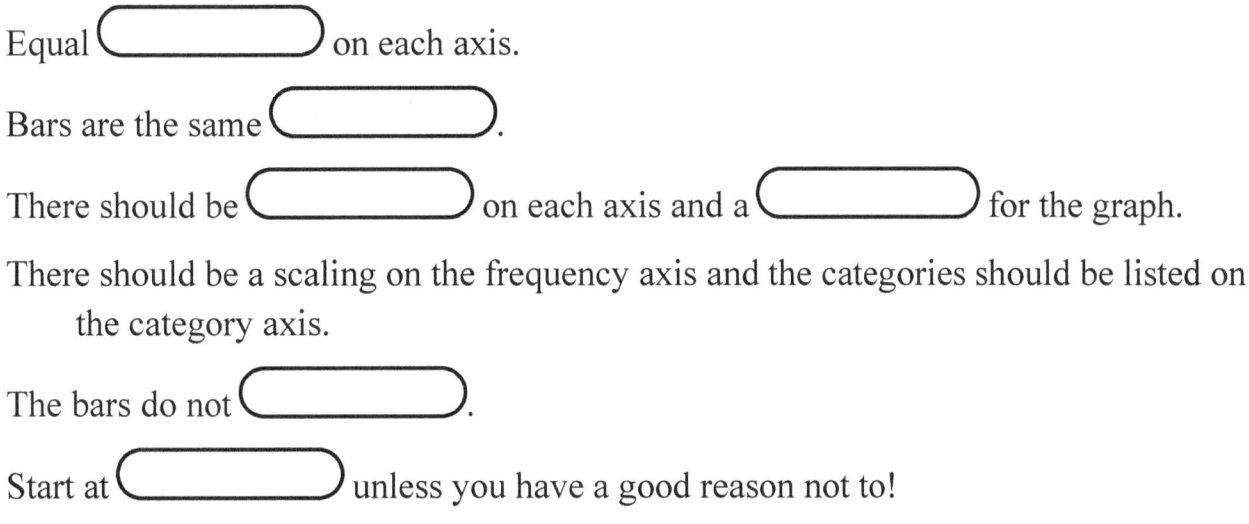 on each axis.

Bars are the same ⬭.

There should be ⬭ on each axis and a ⬭ for the graph.

There should be a scaling on the frequency axis and the categories should be listed on the category axis.

The bars do not ⬭.

Start at ⬭ unless you have a good reason not to!

Relative Frequency Bar graphs

You can also draw a bar graph using relative frequency on the vertical axis. This is useful when you want to compare two samples with different ⬭ sizes. The relative frequency graph and the frequency graph should look the same, except for the scaling on the frequency axis.

## Pie Charts

These became really popular when inexpensive computer graphics became available

## Type of Car Driven

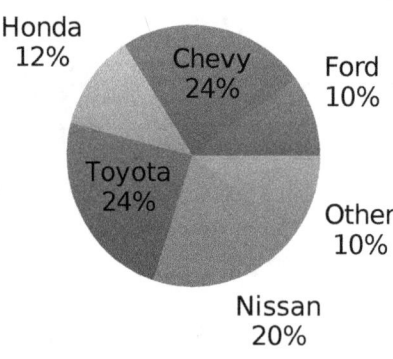

Be wary of colors

Colors can make graphics stunning, but many are ultimately printed in black and white, like this workbook.

Avoid "color legends" that make it very hard to match up $\bigcirc$ to their areas.

## Pareto Charts

A third type of qualitative data graph is a **Pareto chart,** which is just a bar chart with the bars sorted with the $\bigcirc$ frequencies on the left. Here is the Pareto chart for our data

## Pareto Chart for Car Type

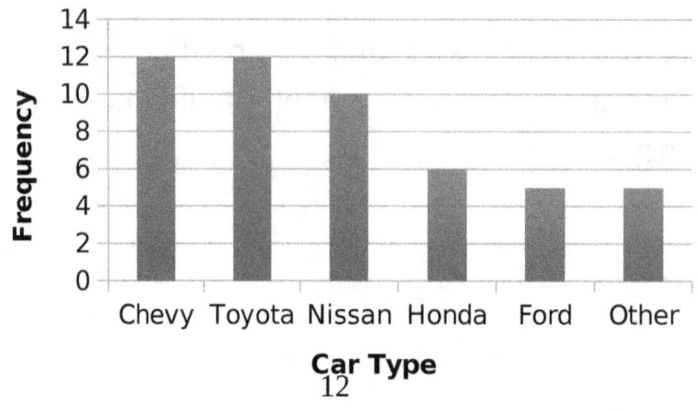

Graphics packages like that in Excel provide lots of different graph types that can put a lot of information on a graph. Use them with care. Make sure they are easy to read.

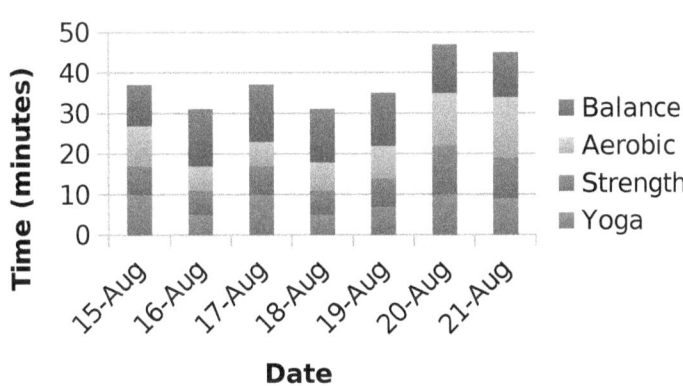

## Section 2.2: Quantitative Data

Frequency distribution Tables

Histograms

Ogives

Stem-and-leaf plots (histograms for typewriters)

Making Histograms

We start by making a frequency distribution.

We divide up the range of data into *frequency classes* or *bins.*

Count how many observations fall into each *bin.*

Now make a histogram – a bar chart of these frequencies. One difference from regular bar charts – we make the bars touch one another.

1)      Find the range = largest value – smallest value

2)      Pick the number of classes to use. Usually the number of classes is between five and twenty. Five classes are used if there are a small number of data points and twenty classes if there are a large number of data points (over 1000 data points). (Note: categories will now be called classes from now on.)

3)      Class width = (range)/(number of classes). Always round up to the next integer (even if the answer is already a whole number go to the next integer). If you don't do this, your last class will not contain your largest data value, and you would have to add another class just for it. If you round up, then your largest data value will fall in the last class, and there are no issues.

4)      Create the classes. Each class has limits that determine which values fall in each class. To find the class limits, set the smallest value as the lower class limit for the first class. Then add the class width to the lower class limit to get the next lower class limit. Repeat until you get all the classes. The upper class limit for a class is one less than the lower limit for the next class.

5)      In order for the classes to actually touch, then one class needs to start where the previous one ends. This is known as the class boundary. To find the class boundaries, subtract 0.5 from the lower class limit and add 0.5 to the upper class limit.

6)      Sometimes it is useful to find the class midpoint. The process is midpoint = (lower limit + upper limit)/2

7)      To figure out the number of data points that fall in each class, go through each data value and see which class boundaries it is between. Utilizing tally marks may be helpful in counting the data values. The frequency for a class is the number of data values that fall in the class.

The previous description is for counting numbers. For measuring (continuous) numbers, we need to use "half-open intervals" to create our bins.

An example of half-open intervals is letter grade assignments:

| A: [94,100]% | A-: [88,94)% | B+: [84,88)% | B: [82,84)% | B-: [79,82)% |
|---|---|---|---|---|
| C+: [75,79)% | C: [72,75)% | C-: [69,72)% | D: [65,69)% | E: below 65% |

## Example

Here is an example of Monthly Rent Data:

| 1500 | 1350 | 350 | 1200 | 850 | 900 |
|---|---|---|---|---|---|
| 1500 | 1150 | 1500 | 900 | 1400 | 1100 |
| 1250 | 600 | 610 | 960 | 890 | 1325 |
| 900 | 800 | 2550 | 495 | 1200 | 690 |

Range: (max – min) 2550 – 350 = 2200

Number of classes (7 in this example)

Class width   2200/7 ≈ 314.286 ≈ 315    (Round up, even if whole)

## Class Limits

Start with the min, and add class width.

## Tally the Data

| Class Limits | Class Boundaries | Class Midpoint | Tally | Frequency |
|---|---|---|---|---|
| 350 – 664 | 349.5 – 664.5 | 507 | \|\|\|\| | 4 |
| 665 – 979 | 664.5 – 979.5 | 822 | 卌 \|\|\| | 8 |
| 980 – 1294 | 979.5 – 1294.5 | 1137 | 卌 | 5 |
| 1295 – 1609 | 1294.5 – 1609.5 | 1452 | 卌 \| | 6 |
| 1610 – 1924 | 1609.5 – 1924.5 | 1767 | | 0 |
| 1925 – 2239 | 1924.5 – 2239.5 | 2082 | | 0 |
| 2240 – 2554 | 2239.5 – 2554.5 | 2397 | \| | 1 |

Draw the histogram

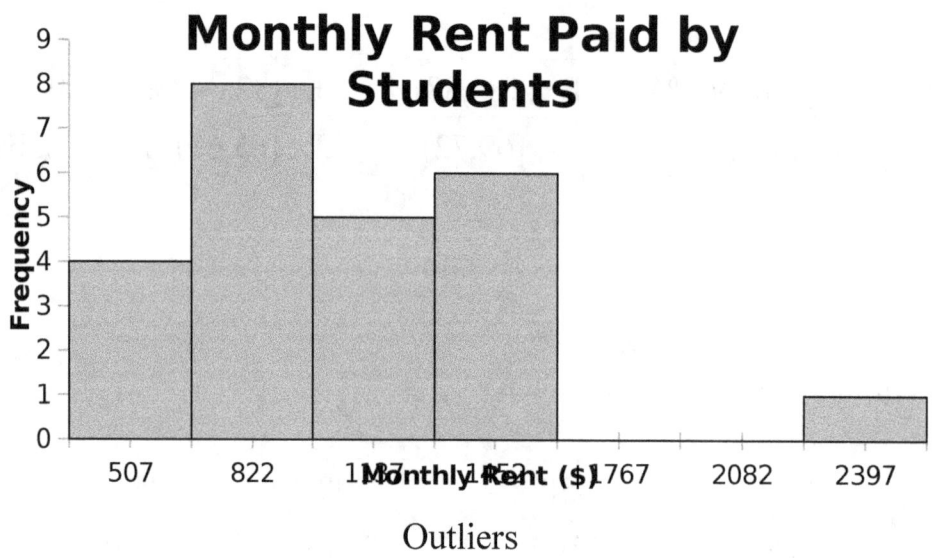

Outliers

An **outlier** is a data value that is ⟨＿＿＿＿＿⟩ from the rest of the values. It may be an ⟨＿＿＿＿＿⟩ value or a ⟨＿＿＿＿＿⟩. It is a data value that should be investigated. In this case, the student lives in a very expensive part of town, thus the value is not a mistake, and is just very unusual. There are other aspects that can be discussed, but first some other concepts need to be introduced.

Relative Frequency histograms

The picture is the same, the y axis is just labeled differently.

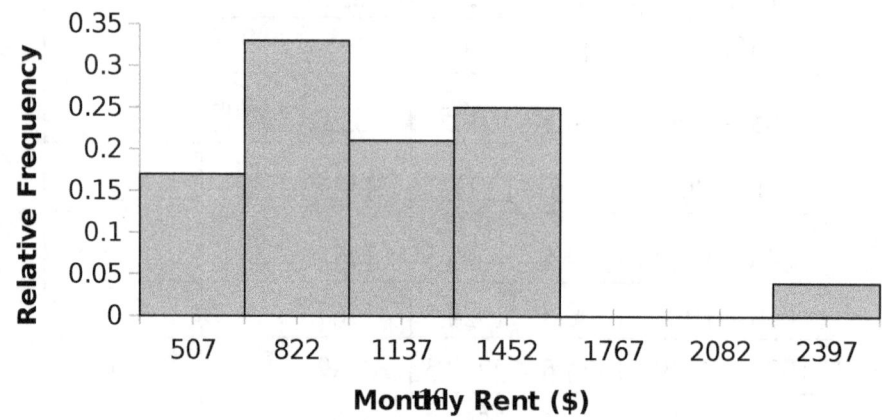

To create a **cumulative frequency distribution**, count the number of data points that are below the upper class boundary, starting with the first class and working up to the top class.  The last upper class boundary should have all of the data points below it.  Also include the number of data points below the lowest class boundary, which is zero.

Ogive: plot of a cumulative frequency distribution.

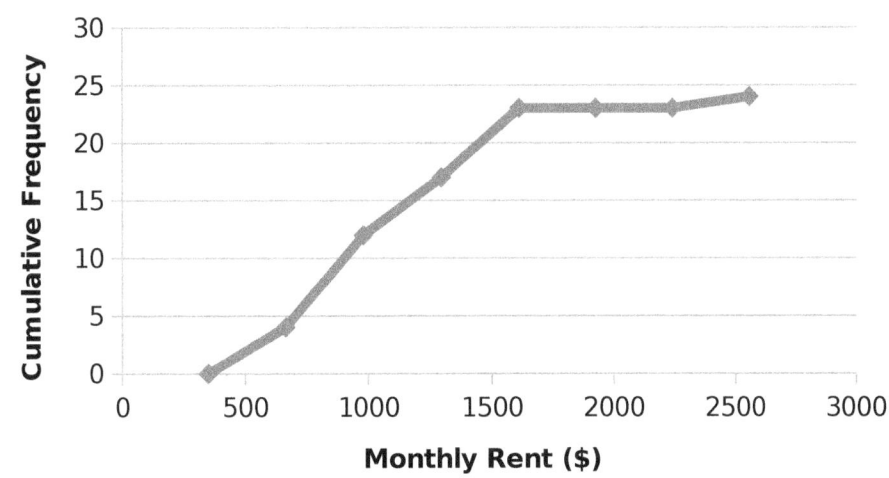

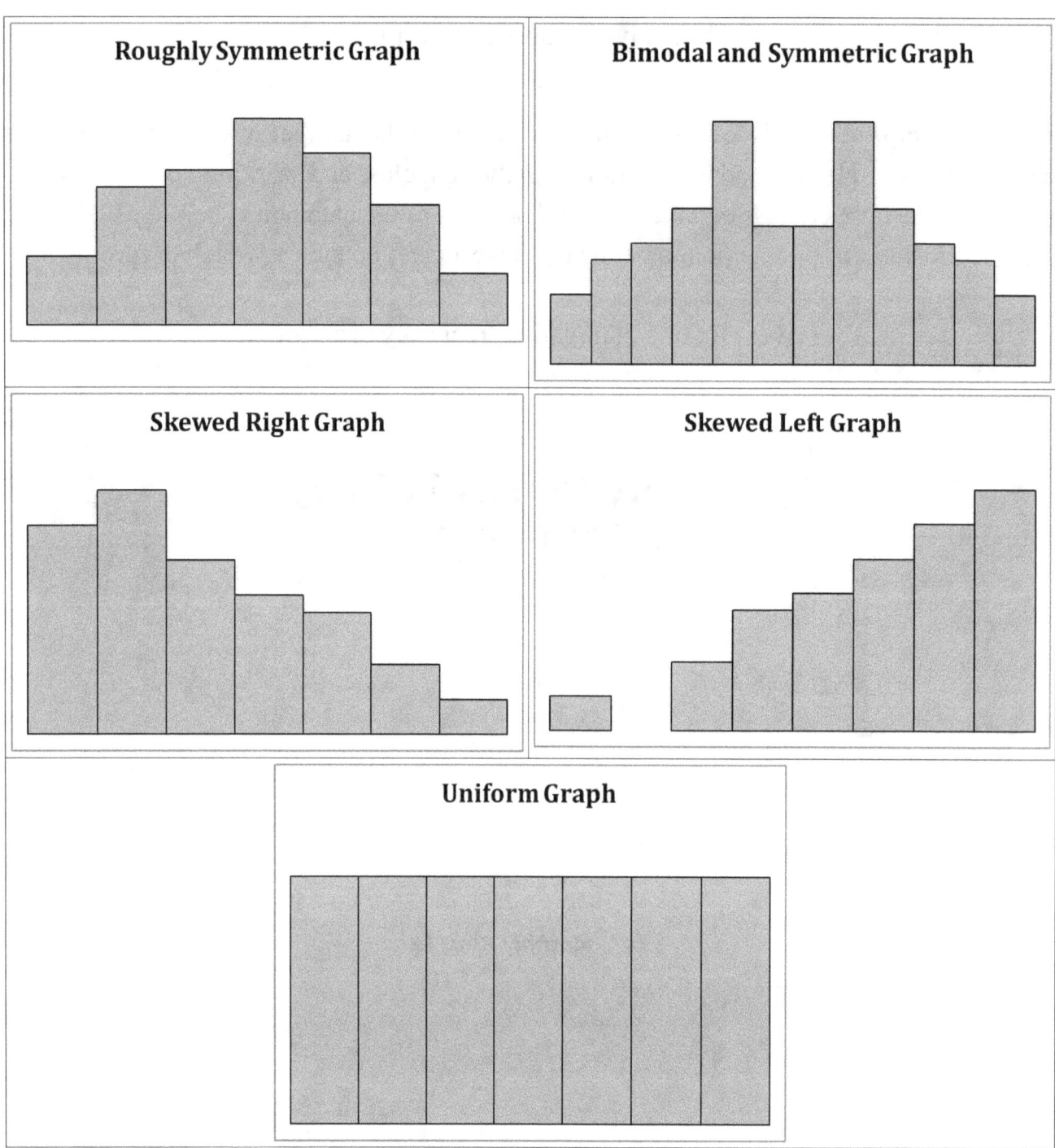

## Section 2.3: Other Graphical Representations of Data

Stem-and-Leaf Plots

   Making histograms with typewriters resulted in the stem-and-leaf plot.

Scatter plots

Time Plots

   You have seen a lot of these.

   Time is always on the $x$ (horizontal) axis

   Shows growth, periodicity, etc.

### Data for a stem-and-leaf

| 62 | 87 | 81 | 69 | 87 | 62 | 45 | 95 | 76 | 76 |
|----|----|----|----|----|----|----|----|----|----|
| 62 | 71 | 65 | 67 | 72 | 80 | 40 | 77 | 87 | 58 |
| 84 | 73 | 93 | 64 | 89 |    |    |    |    |    |

Graph a score of 62:

```
4 |
5 |
6 | 2
7 |
8 |
9 |
```

Complete the rest of the graph:

```
4 | 5 0
5 | 8
6 | 2 9 2 2 5 7 4
7 | 6 6 1 2 7 3
8 | 7 1 7 0 7 4 9
9 | 5 3
```

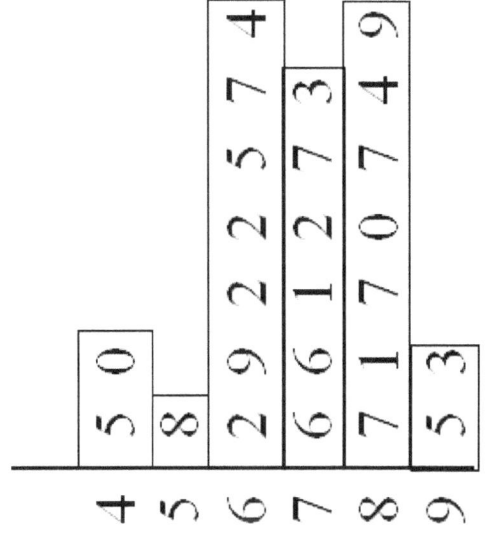

| Elevation (in feet) | 7000 | 4000 | 6000 | 3000 | 7000 | 4500 | 5000 |
|---|---|---|---|---|---|---|---|
| Temperature (°F) | 50 | 60 | 48 | 70 | 55 | 55 | 60 |

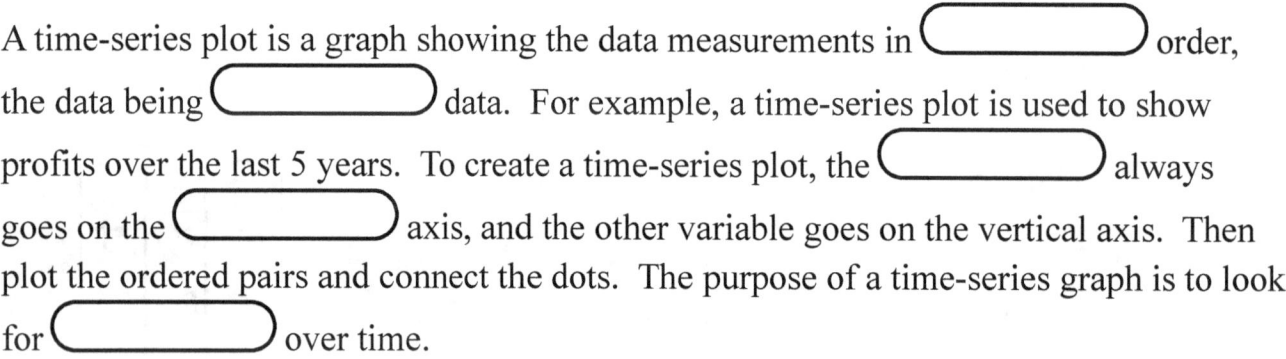

**Temperature versus Elevation**

Time-Series

A time-series plot is a graph showing the data measurements in ⬭ order, the data being ⬭ data. For example, a time-series plot is used to show profits over the last 5 years. To create a time-series plot, the ⬭ always goes on the ⬭ axis, and the other variable goes on the vertical axis. Then plot the ordered pairs and connect the dots. The purpose of a time-series graph is to look for ⬭ over time.

Example

| Time (months) | 0 | 1 | 2 | 3 | 4 |
|---|---|---|---|---|---|
| Weight (pounds) | 200 | 195 | 192 | 193 | 190 |

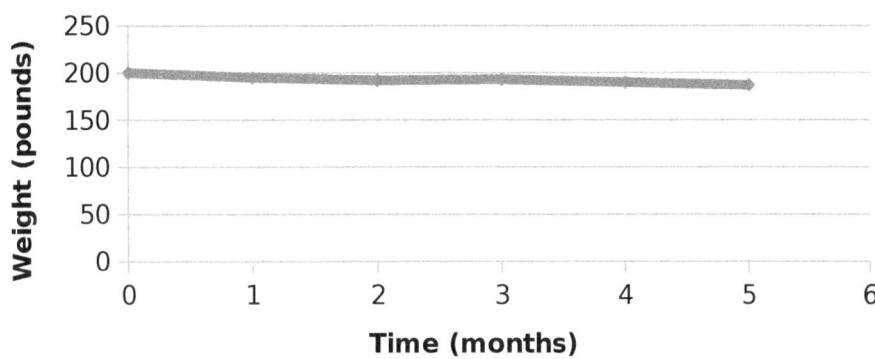

Is this exaggerated?

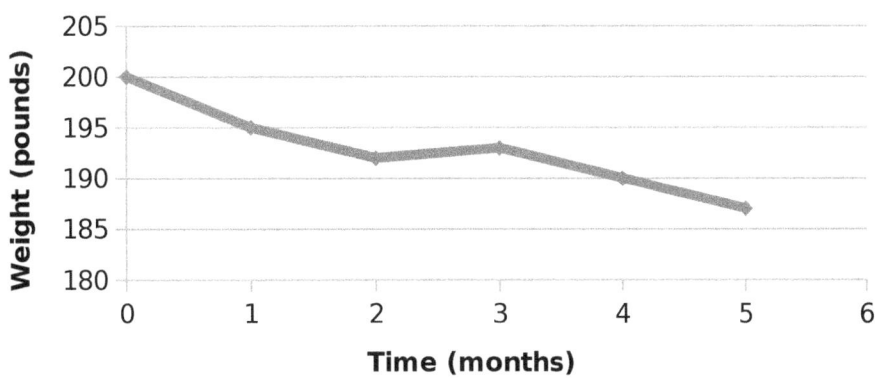

# Chapter 3
## Numerical Descriptions of Data

### 3.1 Measures of Center

- Mode

- Mean

- Median

## Mode

- Most frequently appearing value, or most common frequency class

- "Peaks" in the distribution

- Need not be all the exact same height or count

- Mostly to recognize bi-modal and multi-modal

- Multi-Modal is often a tip-off that different types of individuals in the data set.

## Mean or Average

- Known as Mean, Arithmetic Mean, and Average.

- Especially useful when the data is roughly symmetrical and without many outliers.

- Can be misleading on very skewed data.

- Consider the average income of people in this classroom.

Population Mean

$$\mu = \frac{\sum x}{N}$$

Where $N$ is the size of the population, $x$ is the random variable, and $\sum x$ is summation of the random variables.

Sample Mean

$$\bar{x} = \frac{\sum x}{n}$$

Where $n$ is the sample size.

Median

- The "halfway point" – roughly half are smaller than this value, have are larger.

- This measure of center is more "resistant" to skewness and $\underline{\hspace{2cm}}$.

- Frequently used for distributions like $\underline{\hspace{2cm}}$ and house cost.

Shapes and measures of center

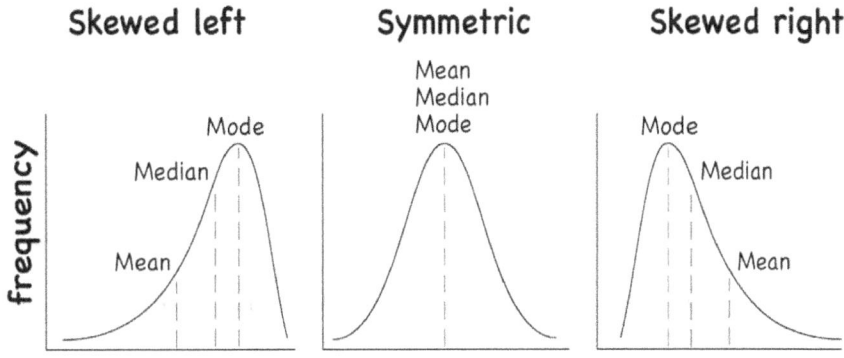

## Weighted Average

- Students are often familiar with this, as grades are usually computed as weighted averages.

- For example:

  - Homework    20%

  - Worksheets   15%

  - Tests       65%

- $w$ is the weight of the data value, $x$

$$\bar{x} = \frac{\sum x w}{\sum w}$$

## Section 3.2: Measures of Spread

Range

    Max-Min

    Obviously, ⟨＿＿＿＿⟩ to outliers

IQR (Interquartile Range)

    Range of ⟨＿＿＿＿⟩ half

    Less ⟨＿＿＿＿⟩ to outliers

Standard Deviation

    Appropriate for symmetric distributions where the mean is a good measure of center.

We are often interested in the source and amount of variation in a data set.

You will find it useful to pause for a moment and ask "What is the source of variation in this data set?"

Often knowing the source of variation can help you understand the shape of the distribution.

# Range

( _____ ) the data makes it easy to find minimum and maximum values.

Probably the most common measure in non-technical situations, but its sensitivity to ( _____ ) can be a problem.

# Standard Deviation

"Average un-averageness".

It is a good description of spread in data that is reasonably free of extreme outliers.

It is ( _____ ) to outliers.

Although not as useful in skewed data sets, it can be calculated for <u>any</u> data set.

Each data point in a data set has a deviation from the mean (average), $\bar{x}$.

The data points can be listed: $x_1, x_2, x_3, x_4, \ldots, x_n$. The "ith $x$" is $x_i$.

The deviation of "ith $x$" is $x_i - \bar{x}$.

By the definition of mean, and the laws of math, if we add up all the deviations, the sum is always ( _____ ).

Squaring the deviations also gets rid of the negatives, so they can be added up on the way to calculating an average. ( _____ ) also has some nice mathematical properties. (smooth graph, rewarding "close")

When dealing with samples is better to divide by one less than the count. This makes for a little bit larger estimate of variability. We call this *variance*.

# Variance

This is the average *squared* deviation.

Think of it as "the last thing you calculate before you take the square root."

# Standard Deviation

Take the square root of the Variance to get the Standard Deviation.

Denoted: $S$ or $S_x$, for the Standard Deviation of the $x$s

$$S_x = \sqrt{\frac{\sum (x_i - \bar{x})^2}{n-1}}$$

Population Standard Deviation

Denoted with a lower case sigma.

$$\sigma = \sqrt{\frac{\sum (x_i - \bar{x})^2}{N}}$$

# Chebyshev's Theorem

For <u>any</u> set of data:

At least 75% of the data fall in the interval from $\mu - 2\sigma$ to $\mu + 2\sigma$.

At least 88.9% of the data fall in the interval from $\mu - 3\sigma$ to $\mu + 3\sigma$.

At least 93.8% of the data fall in the interval from $\mu - 4\sigma$ to $\mu + 4\sigma$.

# Z-Score and un-usualness

How far is a data point from the mean?

If the data value is outside two standard deviations of the mean, either above or below, then the number is *uncommon*.

## 3.3 Ranking

Percentile: ranks by $100^{ths}$

Decile: ranks by $10^{ths}$

Quartile: ranks by $4^{ths}$

Quintiles: ranks by $5^{ths}$

Percentile: The **kth percentile** is the data value that has k% of the data at or below that value.

For example, "The 1%" refers to the highest-earning 1% of Americans. The $99^{th}$ percentile is about $400,000

# Quartiles

Sort the data in increasing order.

Find the median, this divides the data list into 2 halves.

Find the median of the data below the median. This value is $Q1$.

Find the median of the data above the median. This value is $Q3$.

# Interquartile Range (IQR)

$IQR = Q3-Q1$  This is another measure of spread.  It is resistant to outliers in much the same way as median is resistant to outliers to express the center.

$Q1-1.5{\times}IQR$ ("Low Fence") and $Q3 + 1.5{\times}IQR$ ("High Fence") provide another definition of "outlier"

# 5-Number Summary - Box and Whisker Plot

These 5 numbers are often used to provide a nice summary of a distribution:

- Minimum
- Q1
- Median
- Q3
- Maximum

A typical box-and-whisker plot

Box-and-Whisker Plot are used to gauge ⬭ and ⬭ of a distribution.

Multiple Box-and-Whisker plots on the same field are convenient to compare two ⬭.

# Chapter 4
## Probability

### 4.1: Empirical Probability

• **Experiment**: an activity that has specific ⬭ that can occur, but it is unknown ⬭ results will occur.

• **Outcomes**: the ⬭ of an experiment

• **Event**: a set of certain outcomes of an ⬭ that you want to have happen

• **Sample Space**: collection of all ⬭ outcomes of the experiment. Usually denoted as SS.

• **Event space**: the set of outcomes that make up an event. The symbol is usually a capital letter.

### Trials for Die Experiment

| $n$ | Number of 6s | Relative Frequency |
|-----|--------------|--------------------|
| 10  | 2            | 0.2                |
| 50  | 6            | 0.12               |
| 100 | 18           | 0.18               |
| 500 | 81           | 0.162              |
| 1000| 163          | 0.163              |

### Experimental Probabilities

• $P(A) = \dfrac{\text{number of times A occures}}{\text{number of times the experiment was repeated}}$

• In our experiment, $\dfrac{163}{1000} = 0.163$

Law of large numbers

• As n increases, the relative frequency ⬭ towards the ⬭ probability value.

Note: probability, relative frequency, percentage, and proportion are all different words for the same concept.

Also, probabilities can be given as percentages, decimals, or fractions.

## Section 4.2: Theoretical Probability

• It is not always feasible to conduct an experiment over and over again, so it would be better to be able to find the probabilities without conducting the experiment. These probabilities are called **Theoretical Probabilities**.

• To be able to do theoretical probabilities, there is an assumption that you need to consider. It is that all of the outcomes in the sample space need to be **equally likely outcomes**. This means that every outcome of the experiment needs to have the same ⬭ of happening.

Theoretical Probabilities

If the outcomes of an experiment are equally likely, then the probability of event A happening is:

$$P(A)=\frac{\text{number of outcomes in event space}}{\text{number of outcomes in sample space}}$$

Example: Flip a pair of coins

• What is the sample space?

• What is the probability of getting exactly one head?

• What is the probability of getting at least one head?

• What is the probability of getting a head and a tail?

• What is the probability of getting a head or a tail?

# Probability Properties

1. $0 \le P(\text{event}) \le 1$

2. If the P(event) = 1, then it will happen and is called the ⬭ event

3. If the P(event) = 0, then it cannot happen and is called the ⬭ event

4. $\sum P(\text{event}) = 1$

### Example: Pull a card from a 52-card deck

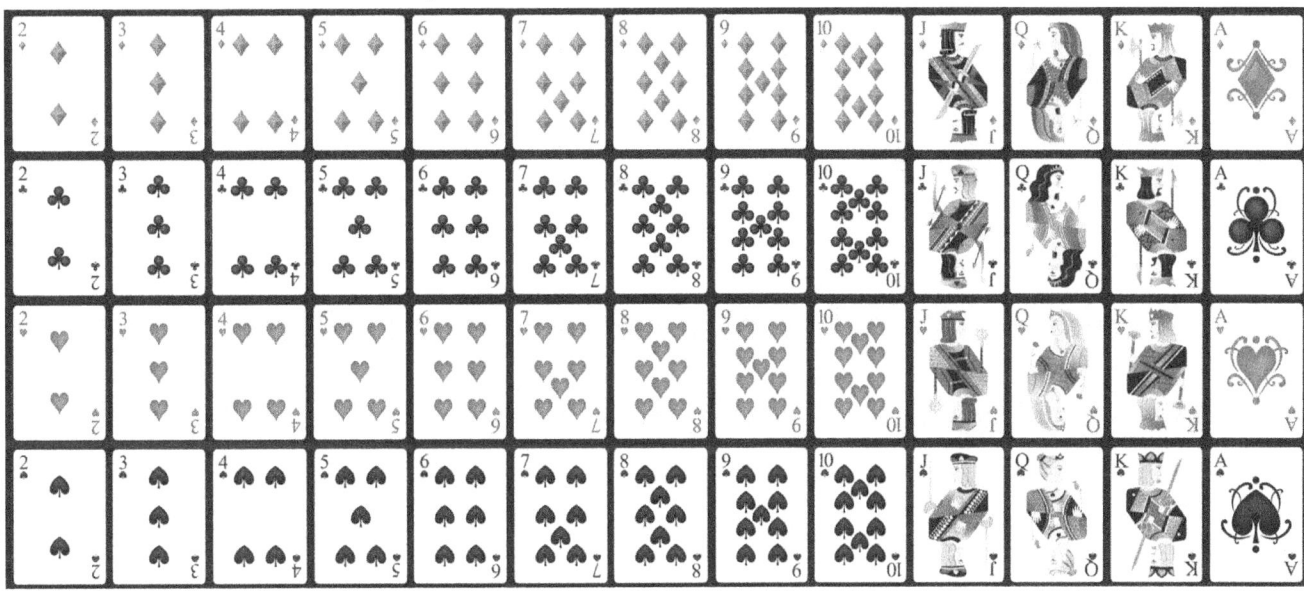

• What is the sample space?

• What is the probability of getting a Spade?

• What is the probability of getting a Jack?

• What is the probability of getting an Ace?

• What is the probability of not getting an Ace?

• What is the probability of getting a Spade and an Ace?

• What is the probability of getting a Spade or an Ace?

• What is the probability of getting a Jack and an Ace?

• What is the probability of getting a Jack or an Ace?

## Complementary events

• If A is an event, the complementary event can be notated: $A^C$ , not A, ~A or $\bar{A}$.

• $P(A) + P(not\ A) = 1$

• $P(not\ A) = 1 - P(A)$

• Sometimes it's a lot ⬭ to calculate P(not A) than P(A)

Shared Birthdays

- What is the probability that two students in this class share a birthday?

- There are a <u>lot</u> of ways this can happen!

- <u>But there is only</u> ⬭ way of it not happening, that is if everybody has a ⬭ birthday.

- What is the probability that at least two friends in a group of eight share a birthday?

- What is the probability that at least two friends in a group of twenty share a birthday?

- What is the probability that at least two friends in a group of thirty share a birthday?

## Mutual Exclusivity

• Two events are mutually exclusive if they ⟨_____⟩ happen at the same time.

• Canonical examples: (rolling a pair of dice)

– **Exclusive: Rolling a pair and rolling a 7**
there is no roll of two dice that totals 7 and has the same number on each die.

– **Not Exclusive: Rolling a pair and rolling an 8**
the roll of 4, 4 totals eight and totals 8 points.

## Independence

• Two events are independent if the fact that one happens does not ⟨_____⟩ the probability of the second happening.

• Canonical examples: (Drawing cards from a 52- card deck. Event A is "Draw a Queen", and Event B is "Draw a Queen")

– **Independent**: draw a card, note whether it is a Queen or not, **put it back in the deck, re-shuffle**, and draw a second card.

– **Not independent**: draw **two cards** out of the deck. The probability of the second card being a Queen changes depending on whether the first card was a Queen.

## Addition Rules

• If two events A and B are mutually exclusive, then:

$$P(A \text{ or } B) = P(A) + P(B), \text{ and } P(A \text{ and } B) = 0$$

• If two events A and B are not mutually exclusive, then:

$$P(A \text{ or } B) = P(A) + P(B) - P(A \text{ and } B)$$

Roll a pair of dice

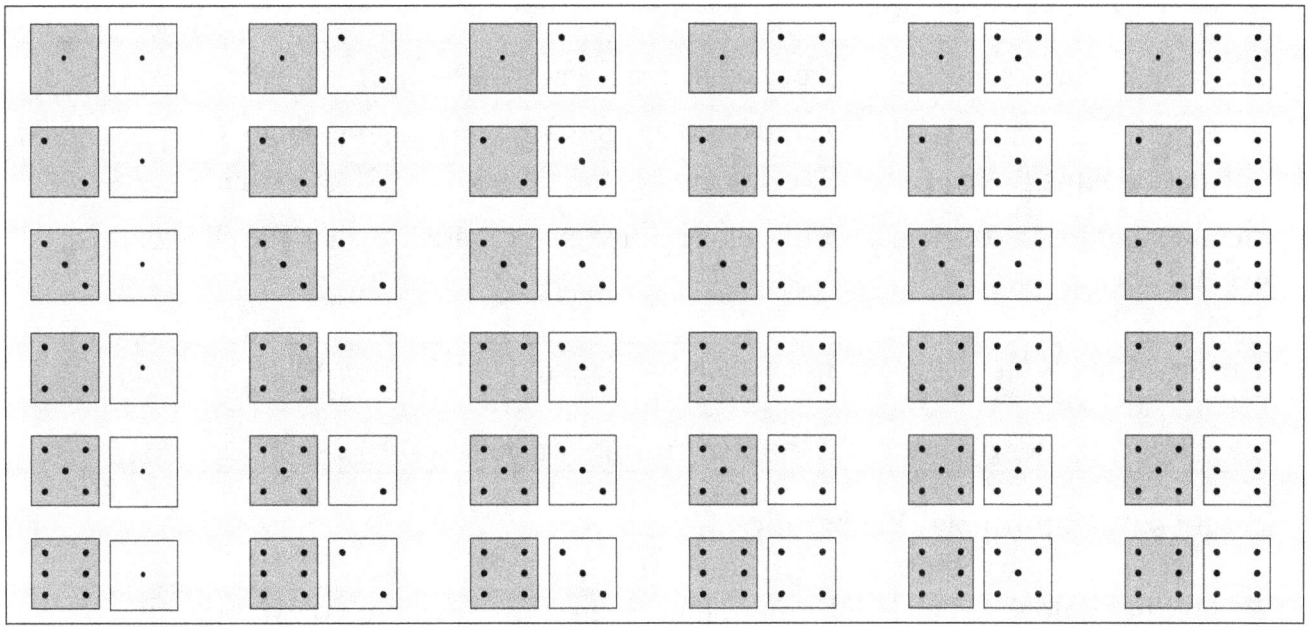

a) What is the sample space?

b) What is the probability of getting a sum of 5?


c) What is the probability of getting the first die a 2?


d) What is the probability of getting a sum of 7?


e) What is the probability of getting a sum of 5 and the first die a 2?


f) What is the probability of getting a sum of 5 or the first die a 2?


g) What is the probability of getting a sum of 5 and sum of 7?


h) What is the probability of getting a sum of 5 or sum of 7?

The **actual odds against** event $A$ occurring are the ratio $P(A^C)/P(A)$, usually expressed in the form $a{:}b$ or $a$ to $b$, where $a$ and $b$ are integers with no common factors.
The **actual odds in favor** event $A$ occurring are the ratio $P(A)/P(A^C)$, which is the reciprocal of the odds against. If the odds against event $A$ are $a{:}b$, then the odds in favor event $A$ are $b{:}a$.

The **payoff odds** against event $A$ occurring are the ratio of the net profit (if you win) to the amount bet.
   payoff odds against event $A$ = (net profit) : (amount bet)

## Section 4.3: Conditional Probability

Probabilities calculated after information is given. This is where you want to find the probability of event A happening after you know that event B has happened. If you know that B has happened, then you don't need to consider the rest of the sample space. You only need the outcomes that make up event B. Event B becomes the new sample space, which is called the **restricted sample space, R.**

### Restricted sample space

If you always write a restricted sample space when doing conditional probabilities and use this as your sample space, you will have no trouble with conditional probabilities. The notation for conditional probabilities is P(A, given B) = P(A|B). The event following the vertical line is always the restricted sample space.

### New Information

One way of looking at this conditional probability issue is "How does this new information cause me to revise my estimate of likelihood (probability)?"

There is a whole branch of statistics known as Bayesian Analysis that deals with this.

**Example:** Suppose you roll two dice. What is the probability of getting a sum of 5, given that the first die is a 2?

**Solution:** Since you know that the first die is a 2, then this is your restricted sample space, so R = {(2,1), (2,2), (2,3), (2,4), (2,5), (2,6)}. Out of this restricted sample space, the way to get a sum of 5 is {(2,3)}. Thus, P(sum of 5 | the first die is a 2) = 1/6

## Probability of rolling a sum of 5?

• When we considered all 36 possible outcomes, 4 of them (1,4), (2,3), (3,2), (4,1) had a total of 5. Without any knowledge of the first die, the probability of getting a 5 is 4/36 = 1/9 about 11.1%

• But, when we know the first die was a 2, it changes the probability to 1/6 or approximately 16.7%.

**Example:** Suppose you roll two dice. What is the probability of getting a sum of 7, given the first die is a 4?

**Solution:** Since you know that the first die is a 4, this is your restricted sample space, so R = {(4,1), (4,2), (4,3), (4,4), (4,5), (4,6)} Out of this restricted sample space, the way to get a sum of 7 is {(4,3)}. Thus P(sum of 7 | first die is a 4) = 1/6

## Probability of rolling a sum of 7?

• When we looked at all 36 outcomes, and 6 of them were 7's, we got the same probability for getting a 7: 6/36 = 1/6

• This means that knowing that the first die is a 4 did not change the probability that the sum is a 7. This added knowledge did not help you in any way. It is as if that information was not given at all.

## Dependent and Independent Events

• In the second case, the events sum of 7 and first die is a 4 are called **independent events**.

• In the first case, the events sum of 5 and first die is a 2 are called **dependent events**.

• Events A and B are considered **independent events** if the fact that one event happens does not change the probability of the other event happening.

If A and B are independent then P(A|B) = P(A), or P(B|A) = P(B).

a) Suppose you roll two dice. Are the events "sum of 7" and "first die is a 3" independent?

b) Suppose you roll two dice. Are the events "sum of 6" and "first die is a 4" independent?

c) Suppose you pick a card from a deck. Are the events "Jack" and "Spade" independent?

d) Suppose you pick a card from a deck. Are the events "Heart" and "Red" card independent?

e) Suppose you flip a coin 50 times and get a head every time, what is the probability of getting a head on the next flip?

## Multiplication Rule

• If two events are dependent, then:

$$P(A \text{ and } B) = P(A) \cdot P(B|A)$$

• If two events are independent, then:

$$P(A \text{ and } B) = P(A) \cdot P(B)$$

• Solving for the conditional probability: $P(B|A) = \dfrac{P(A \text{ and } B)}{P(A)}$

## Multiplication Rule Examples

a) Suppose you pick three cards from a deck, what is the probability that they are all Queens if the cards are not replaced after they are picked?

b) Suppose you pick three cards from a deck, what is the probability that they are all Queens if the cards are replaced after they are picked and before the next card is picked?

## Two-Way Table: Leprosy Cases

| WHO Region | World Bank Income Group | | | | Row Total |
|---|---|---|---|---|---|
| | High Income | Upper Middle Income | Lower Middle Income | Low Income | |
| Americas | 174 | 36028 | 615 | 0 | 36817 |
| Eastern Mediterranean | 54 | 6 | 1883 | 604 | 2547 |
| Europe | 10 | 0 | 0 | 0 | 10 |
| Western Pacific | 26 | 216 | 3689 | 1155 | 5086 |
| Africa | 0 | 39 | 1986 | 15928 | 17953 |
| South-East Asia | 0 | 0 | 149896 | 10236 | 160132 |
| Column Total | 264 | 36289 | 158069 | 27923 | 222545 |

a) Find the probability that a person with leprosy is from the Americas.

b) Find the probability that a person with leprosy is from a high-income country.

c) Find the probability that a person with leprosy is from the Americas and a high-income country.

d) Find the probability that a person with leprosy is from a high-income country, given they are from the Americas.

e) Find the probability that a person with leprosy is from a low-income country.

f) Find the probability that a person with leprosy is from Africa.

g) Find the probability that a person with leprosy is from Africa and a low-income country.

h) Find the probability that a person with leprosy is from Africa, given they are from a low-income country.

i) Are the events that a person with leprosy is from "Africa" and "low-income country" independent events? Why or why not?

j) Are the events that a person with leprosy is from "Americas" and "high-income country" independent events? Why or why not?

# Bayes Theorem

• Begin with:  $P(A \text{ and } B) = P(A) \cdot P(B|A)$

• Goal: calculate P(H|E). This is the new probability of a Hypothesis H, given new evidence E.

• Begin by re-naming:  $P(E \text{ and } H) = P(E) \cdot P(H|E)$

• Note that "and" is symmetric:

$P(E \text{ and } H) = P(H \text{ and } E)$
$P(E) \cdot P(H|E) = P(H) \cdot P(E|H)$

• Solve for P(H|E), the probability that our hypothesis is true, H, given the evidence, E:

$$P(H|E) = \frac{P(H) \cdot P(E|H)}{P(E)}$$

A much more useful form of Bayes Theorem requires use to eliminate the P(E) term. We do this with two substitutions:

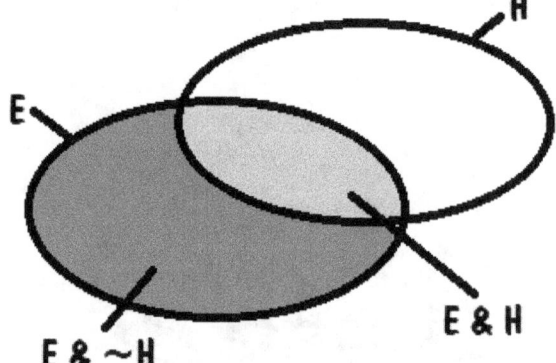

1) $P(E)=P(E \text{ and } H)+P(E \text{ and } \bar{H})$

Notice that
$P(E \text{ and } H)=P(H \text{ and } E)=P(H) \cdot P(E|H)$ and

$P(E \text{ and } \bar{H})=P(\bar{H} \text{ and } E)=P(\bar{H}) \cdot P(E|\bar{H})$ .

Thus, $P(E)=P(E \text{ and } H)+P(E \text{ and } \bar{H})=P(H) \cdot P(E|H)+P(\bar{H}) \cdot P(E|\bar{H})$

$$P(H|E)=\frac{P(H) \cdot P(E|H)}{P(H) \cdot P(E|H)+P(\bar{H}) \cdot P(E|\bar{H})}$$

2) $P(\bar{H})=1-P(H)$

Thus, $P(H|E)=\dfrac{P(H) \cdot P(E|H)}{P(H) \cdot P(E|H)+(1-P(H)) \cdot P(E|\bar{H})}$

## The Three Terms of Bayes Theorem:

$P(H|E)$ can be calculated from only three terms!

• P(H) is called the Prior Probability of H, or just the Prior. Sometimes called the base rate. This is the best estimate of the probability of the hypothesis before considering the new evidence E.

• P(E | H) is the probability of getting the evidence we got under the assumption that H is true.

• P(E | ~H) is the probability of getting the evidence we got under the assumption that H is false. Note that we can use the compliment rule to say:
• P(E | ~H) = 1 – P(~E | ~H).

- **Confirmation bias**: only focusing on P(E|H)

If P(E|H) is high, we may say that the evidence is consistent with our hypothesis!

However, if we ignore P(E | ~H), we may fail to learn that evidence is also consistent with the hypothesis being false. That is, there could be other explanations for the evidence, even if the hypothesis is not correct.

- **Base Rate**: Forgetting to factor in that base rate.

*"The test is 99% accurate, you have positive result, hence there is a 99% chance you have the disease."*

## Apply to medical diagnosis

- **Sensitivity**: Assume a test is 99% accurate when the patient has the disease. This is called the "true positive rate". This means that if the patient has the disease, 99% of the time the test will be positive, indicating that the patient has the disease. Let "H" be the hypothesis that the patient has the disease. Let "E" be the evidence that the test is positive. In this example, P(E | H) = 0.99. We could also say the false negative rate is 1% and P(~E | H) = 0.01.

- **Specificity**: Assume that the test is 98% accurate in the true negative rate, which means that if the patient does not have the disease, 98% of the time the test will be negative. In this example, P(~E | ~H) = 0.98. We could also say the false positive rate is 2% and P(E | ~H) = 0.02.

- **Base Rate**: Assume that 1/1000 of people in the population have the disease. This is the *prior*. In this example, P(H) = 0.001.

Bayes Answers the Question "Do I have the disease?"

$$P(H|E) = \frac{P(H) \cdot P(E|H)}{P(H) \cdot P(E|H) + (1 - P(H)) \cdot P(E|\bar{H})} \quad \text{or}$$

$$P(H|E) = \frac{P(H) \cdot P(E|H)}{P(H) \cdot P(E|H) + (1 - P(H)) \cdot (1 - P(\bar{E}|\bar{H}))}$$

$$\frac{0.001 \cdot 0.99}{0.001 \cdot 0.99 + (1 - 0.001) \cdot (1 - 0.98)} \approx 4.7\%$$

### Population table approach

|  | diseased | healthy | total |
|---|---|---|---|
|  | 100 | 99900 | 100000 |
| Tests + | 99 | 1998 | 2097 |
| Tests - | 1 | 97902 | 97903 |

| P(H | E) | 99/2097 = | 0.04721 |
|---|---|---|

### History and Bayes

• For a long time, there was a great controversy between "Bayesians" and "Frequentests" among statisticians.

• Lately, a consensus seems to be developing that both approaches are just different ways of viewing the world, and are in fact fully compatible.

### Summary of logic to probability rules

• **NOT**:
P(not A) = 1 – P(A)

• **OR**:
P(A or B) = P(A) + P(B) (if P(A and B) =0)

P(A or B) = P(A) +P(B) – P(A and B)

• **AND**:
P(A and B) = P(A) × P(B) (if independent)

P(A and B) = P(A) × P(B | A) (if the probability of B depends on whether A occurred)

44

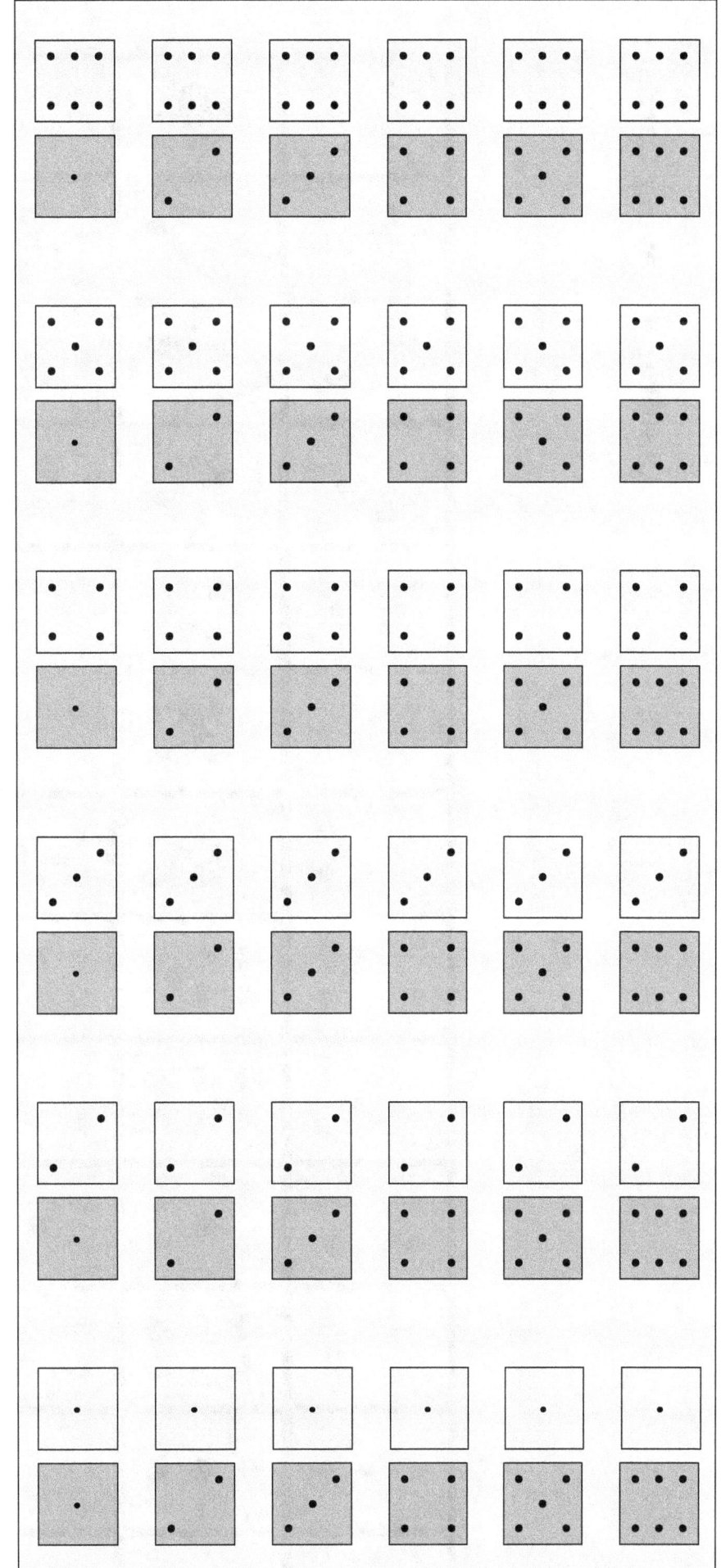

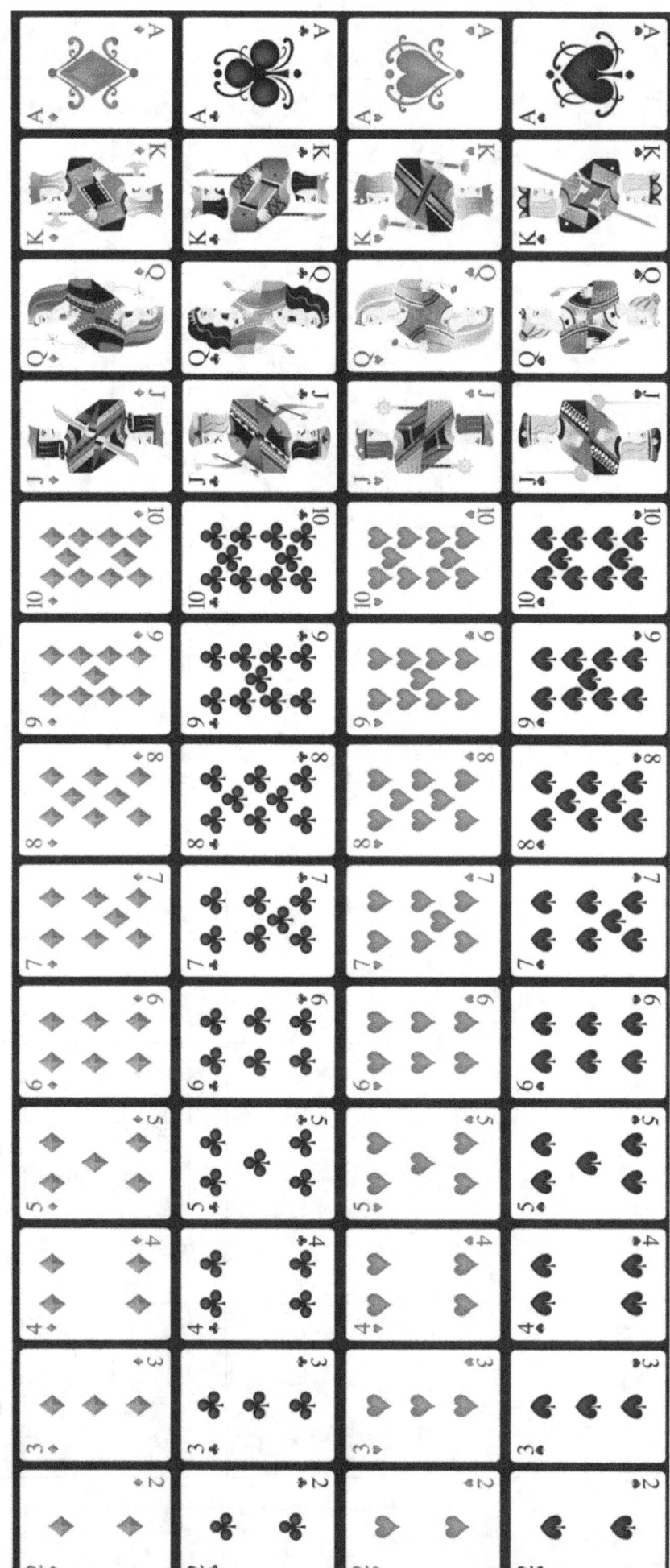

# Chapter 5
## Discrete Probability Distributions

### Discrete vs. Continuous

• As a reminder, the random variable from now on, is represented by the letter $x$ and it represents a quantitative (numerical) variable that is ( _____ ) or ( _____ ) in an ( _____ ) or ( _____ ).

• Also remember there are different types of quantitative variables, called discrete or continuous. What is the difference between discrete and continuous data? ( _____ ) data can only take on particular values in a range. ( _____ ) data can take on any value in a range. Discrete data usually arises from ( _____ ) while continuous data usually arises from ( _____ ).

Size of Household as discrete probability distribution

| Household Size from U.S. Census of 2010 | | | | | | | |
|---|---|---|---|---|---|---|---|
| Size of household | 1 | 2 | 3 | 4 | 5 | 6 | 7 or more |
| Probability | 26.7% | 33.6% | 15.8% | 13.7% | 6.3% | 2.4% | 1.4% |

Mean and St. Deviation of a discrete probability distribution
• The mean is given by:
$$\mu = \sum x P(x)$$
• The mean is also known as **expected value**.

• The standard deviation is given by:
$$\sigma = \sqrt{\sum (x - \mu)^2 P(x)}$$

# Rare Event Rule

If, under a given assumption, the probability of a particular observed event is extremely small, then you can conclude that the assumption is probably not correct.

Determining if an event is unusual If you are looking at $x$ successes among $n$ trials, and the P($x$ or more successes) < 0.05, then you can consider the $x$ an unusually high number of successes. Another way to think of this is if the probability of at least $x$ successes among $n$ trials is less than 0.05, then the event is considered unusual.

If you are looking at $x$ successes among n trials, and the P($x$ or fewer successes) < 0.05, then you can consider the $x$ an unusually low number of successes. Another way to think of this is if the probability of at most $x$ successes among $n$ trials is less than 0.05, then the event $x$ is considered unusual.

Unusual families?

a) Is it unusual for a household to have six people in the family?

b) If you did come upon many families that had six people in the family, what would you think?

c) Is it unusual for a household to have four people in the family?

d) If you did come upon a family that has four people in it, what would you think?

## Section 5.2: Binomial Probability Distribution

1) **Fixed number** of trials, $n$, which means that the experiment is repeated a
( _____ ) number of times.

2) The $n$ trials are **independent**, which means that what happens on one trial does not
( _____ ) the outcomes of other trials.

3) There are **only two outcomes**, which are called ( _____ ) and ( _____ ).

4) The probability of a success **does not change** from trial to trial, where $p$ is the
probability of ( _____ ) and $q$ is the probability of ( _____ ), $q = 1 - p$.

Binomial formula for probability

$$P(x=r) = {}_nC_r\, p^r q^{n-r}, \text{ where } {}_nC_r = \frac{n!}{r!(n-r)!}$$

Note: $p^r q^{n-r}$ is the probability of getting $r$ successes among $n - r$ failures and ${}_nC_r$, or
C(n,r), is the number of ways these $n$ number can be rearranged.

"At Least" and "At Most"

Consider a binomial with 17 possibilities:
0 1 2 3 4 5 6 7 8 9 10 11 12 13 14 15 16 17

Ex: "At least 11" would be these:
0 1 2 3 4 5 6 7 8 9 10 <u>11 12 13 14 15 16 17</u>

Ex.: "At most 10" would be:
<u>0 1 2 3 4 5 6 7 8 9 10</u> 11 12 13 14 15 16 17

"At least 11" is the complement of "at most 10".


Ex.: When looking at a person's eye color, it turns out that 1% of people in the world has green eyes. Consider a group of 20 people.

a) State the random variable.

b) Argue that this is a binomial experiment



Find the probability that:

c) None have green eyes.

d) Nine have green eyes.

e) At most three have green eyes.

f) At most two have green eyes.

g) At least four have green eyes.

h) In Europe, four people out of twenty have green eyes. Is this unusual?

Ex.: Autism

1 in 88 have autism. Group of 10 children.

a) State the random variable

b) Argue that this is a binomial experiment

Find the probability that:

c) None have autism.

d) Seven have autism.

e) At least five have autism.

f) At most two have autism.

g) Suppose five children out of ten have autism. Is this unusual? What does that tell you?

## Section 5.3: Mean and Standard Deviation of Binomial Distribution

- $n$ is number of trials, $p$ is the probability of success, and $q = 1 - p$

- Mean $\mu = n\,p$

- Standard Deviation $\sigma = \sqrt{npq}$

Ex.: 20 people, when 1% have green eyes

a) State the random variable.

b) Argue that this is a binomial experiment

c) Write the probability distribution.
d) Draw a histogram.

e) Find the mean.

f) Find variance.

g) Find the standard deviation.

| x | P(x) |
|---|---|
| 0 | |
| 1 | |
| 2 | |
| 3 | |
| 4 | |
| 5 | |
| 6 | |
| 7 | |
| 8 | |
| 9 | |
| 10 | |
| 11 | |
| 12 | |
| 13 | |
| 14 | |
| 15 | |
| 16 | |
| 17 | |
| 18 | |
| 19 | |
| 20 | |

# Chapter 6
## Continuous Probability Distributions

### 6.1 Uniform Distribution

If you have a situation where the probability is always the same, then this is known as a uniform distribution. An example would be waiting for a train. The trains on the Blue and Green Lines for the Regional Transit Authority (RTA) in Cleveland, OH, have a waiting time during peak hours of ten minutes ("2012 annual report," 2012). If you are waiting for a train, you have anywhere from zero minutes to $\bigcirc$ minutes to wait. Your probability of having to wait $\bigcirc$ number of minutes in that interval is the $\bigcirc$.

Uniform Distribution Graph

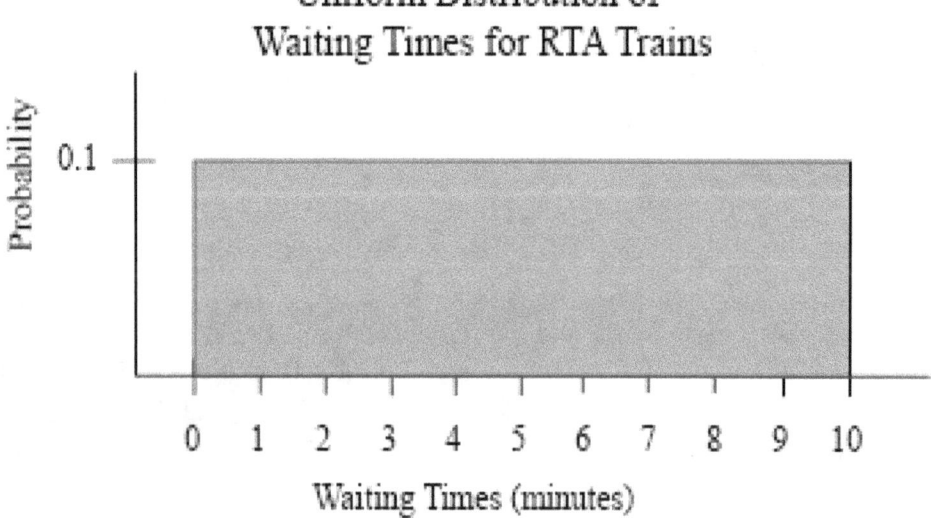

Probability of an interval

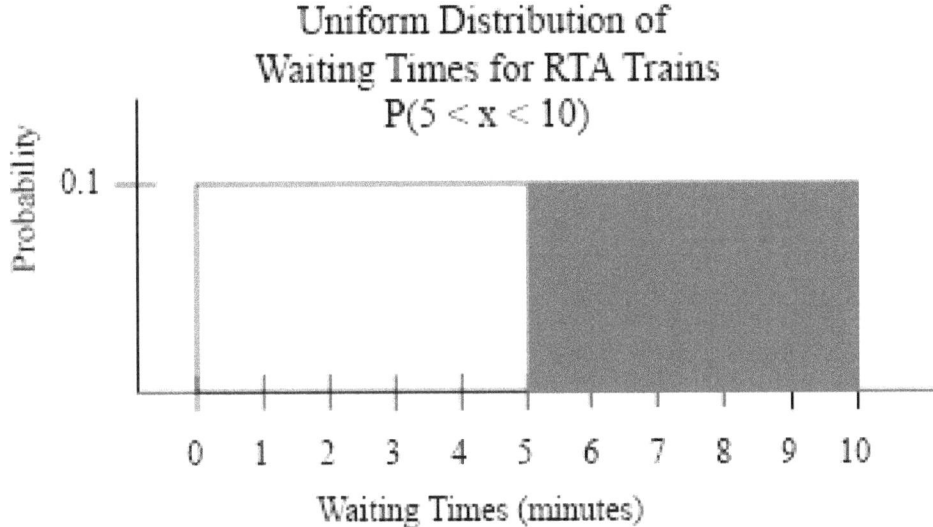

Questions about a uniform distribution: the RTA wait time

a) State the random variable.

b) Find the probability that you have to wait between four and six minutes for a train.

c) Find the probability that you have to wait between three and seven minutes for a train.

d) Find the probability that you have to wait between zero and ten minutes for a train.

e) Find the probability of waiting exactly five minutes.

Why study uniform distributions?

• Computer random number generators are used for a lot of things. They usually generate a ⬭ number in the interval [0 , 1)

• They introduce the idea of probability as an ⬭ under a line or curve.

• They can be used to construct "toy" problems for a statistics class.

• For example, the sum of two uniform random numbers makes a slightly more interesting (but still calculable) geometry for areas.

## 6.2: Graphs of the Normal Distribution

"Normal" has a special meaning here. Many real life problems produce a continuous probability distribution (histogram) that is a symmetric, ( ), and ( ). For example: height, blood pressure, IQ, and cholesterol level. However, not every bell shaped curve is a ( ) curve. In a normal curve, there is a specific relationship between its "height" and its "width."

Typical Graph of a Normal Curve

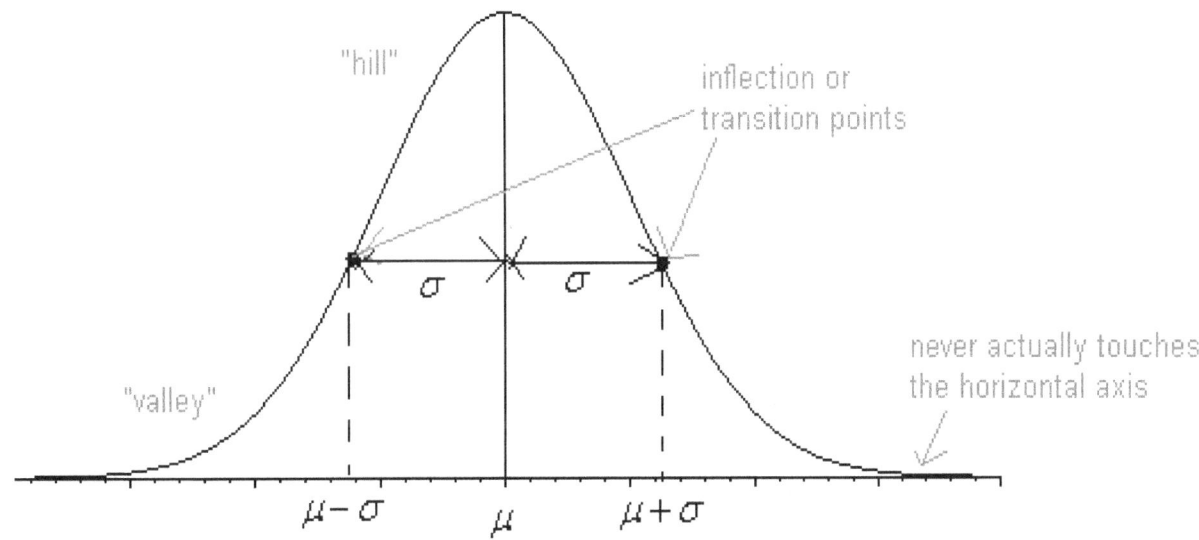

Characteristics of a Normal Curve

• The center, or the highest point, is at the population mean, $\mu$.
• The transition points (inflection points) are the places where the curve changes from a "hill" to a "valley". The distance from the mean to the ( )( ) is one standard deviation, $\sigma$.
• The area under the whole curve is exactly ( ). Therefore, the area under the half below (or above) the mean is ( ).

Comparing "≤" with "<"
• In terms of probabilities, there is no difference between $a \le x \le b$ and $a < x < b$.

• For example, the probability of being taller than 6 feet tall, and the probability of being 6 feet tall or taller, is exactly the same. This is true because no one is $\underline{\hphantom{exactly}}$ 6 feet tall!

Probability of an Event is an Area

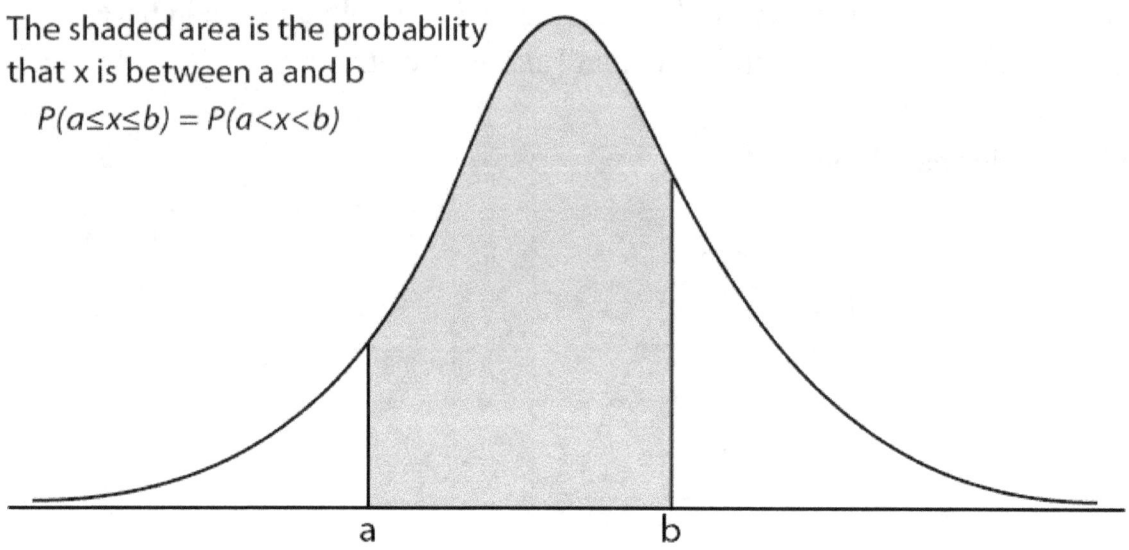

The shaded area is the probability that x is between a and b

$P(a \leq x \leq b) = P(a < x < b)$

a　　　　b

Empirical Rule

**The Empirical Rule** for any **normal** distribution:

• Approximately 68% of the data is within $\bigcirc$ standard deviation of the mean.

• Approximately 95% of the data is within $\bigcirc$ standard deviations of the mean.

• Approximately 99.7% of the data is within $\bigcirc$ standard deviations of the mean.

(This like Chebyshev's rule, but only true for normal distributions.)

Empirical Rule

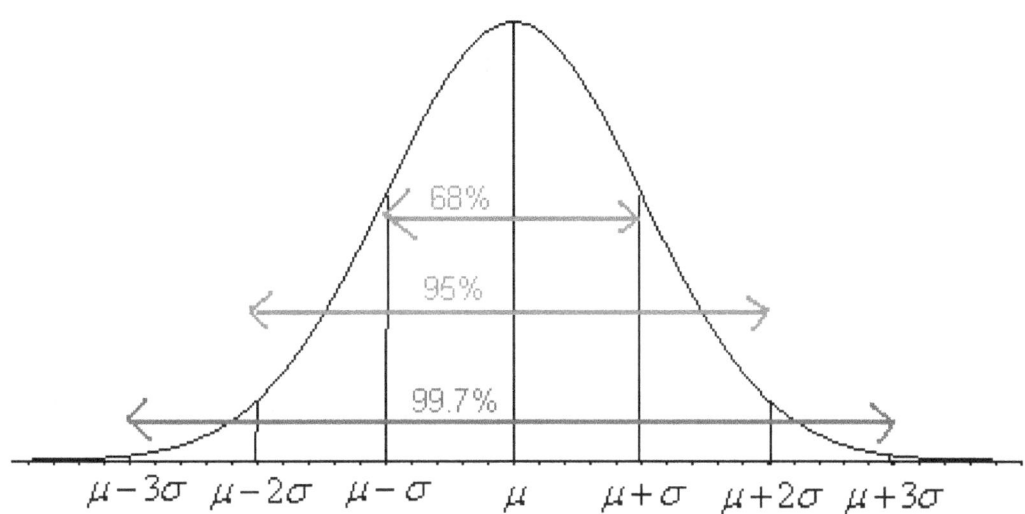

*z*-score

• The *z*-score is normally distributed, with a mean of 0 and a standard deviation of 1. It is known as the standard normal curve.

• Before modern technology, we had to calculate a z-score to look-up in a table.

Standard Normal Curve

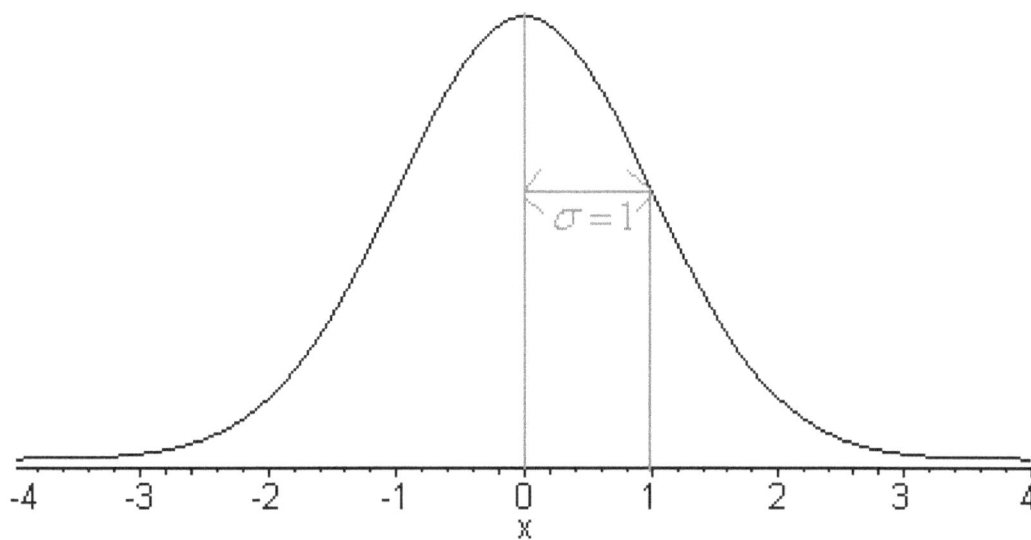

## Normal Distribution Calculator

The Normal Distribution Calculator preforms probability calculations and provides a graph.

You can start exploring the normal curve by finding areas under a region of the curve. For example, a problem may ask you to find P($z$ > 2.0), the probability that $z$ is greater than 2.0.

A $z$-value is from a standard normal distribution. For a standard normal distribution we know that the mean is zero, $\mu = 0$, and that the standard deviation is one, $\sigma = 1$. (If we do not know the population standard deviation, $\sigma$, and instead know the sample standard deviation, $S_x$, we will use the Gosset Student T Distribution Calculator.) Thus, P($z$ > 2.0) asks how likely is it that a randomly chosen value in a normal distribution will be more than 2 standard deviations above the mean.

To find P($z$ > 2.0), we are asked to calculate a probability, and enter 0 for mu and 1 for sigma. Since we are being asked for a probability, we select the box "Find Probability." Since we are asked for a $z$-value greater than 2, we are select the box "Right Tail" to find the probability that a randomly selected value is to the right. When the "Lower Bound" field appears, we enter 2. You may need to click away from the field, press Enter on the keyboard or ← in the keypad to calculate.

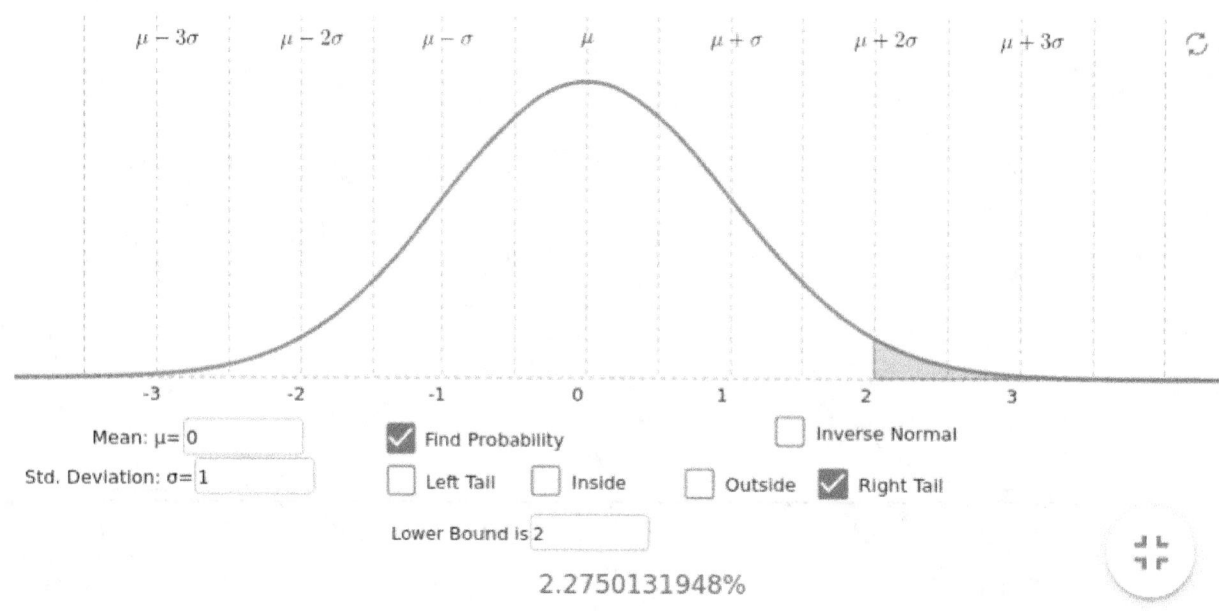

• If we are finding the probability that $z$ is less than the boundary, we calculate the area to the left of the boundary by selecting "Left Tail." Similarly, if $z$ is greater than the boundary, we calculate the area to the right by selecting "Right Tail."

• If you have two boundaries, you are either interested in the inside area between the boundaries by selecting "Inside," or the area outside by selecting "Outside."

Example: Male Heights

In this example, we'll explore real-world numbers. For example, college males have heights normally distributed with a mean of about 70 inches and a standard deviation of 2.8 inches. What is the probability that a randomly chosen blind date is over 6 feet tall?

We know both that the distribution is normal and the population standard deviation, so we can compute the probability using the use the Normal Distribution Calculator. We notice that the units are different, inches and feet. To avoid division introducing more decimals by converting inches to feet, we use multiplication to convert feet into inches. We know that the distribution is normal, $\mu = 70$, $\sigma = 2.8$, and are asked to find $P(x > 72)$.

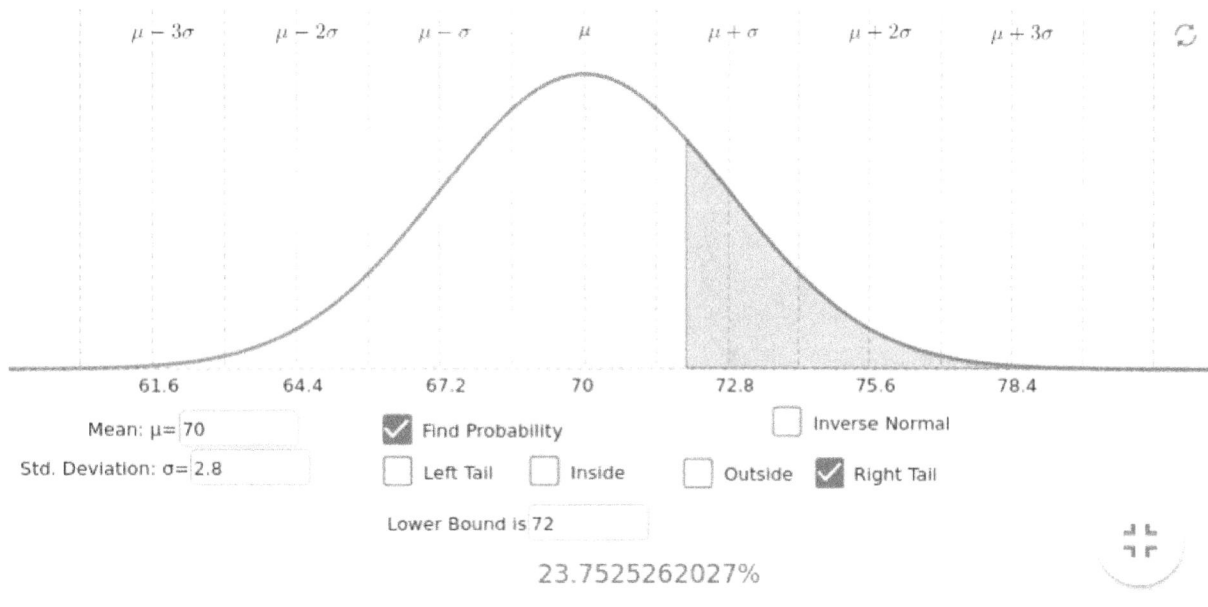

Assuming that a blind date is sufficiently random, $P(x > 72) \approx 0.2375$.

The Normal Distribution Calculator not only allows you to find the probability that a randomly selected value from a distribution is in the specified region by selecting "Find Probability," it also allows you to work backwards, from the area/probability to find the boundaries, by selecting "Inverse Normal."

Sometimes we know an area/probability, and we want to know what $x$ or $z$ value traps that area to the left or right. Sometimes we even have an "inside" area, centered around the mean, and we want to know what values trap that area. If these are $z$-values, we refer to these as $Z^*$ or $Z_C$ or $Z_{/2}$. These are **critical values**.

Example: Confirm that the Empirical Rule for 95% of the data is within $2\sigma$.

The Empirical Rule for 95% of the data is a simple way to recall that approximately 95% of the data in a normal distribution is within 2 standard deviations of the mean. We can use a standard normal distribution, where $\mu=0$ and $\sigma=1$. We want to find the probability that a randomly selected value will be inside 2 standard deviations. That is, we want the probability of being more than -2 and less than 2, $P(-2 < z < 2)$. Using the Normal Distribution Calculator, once we select "Inside" we can enter the boundaries.

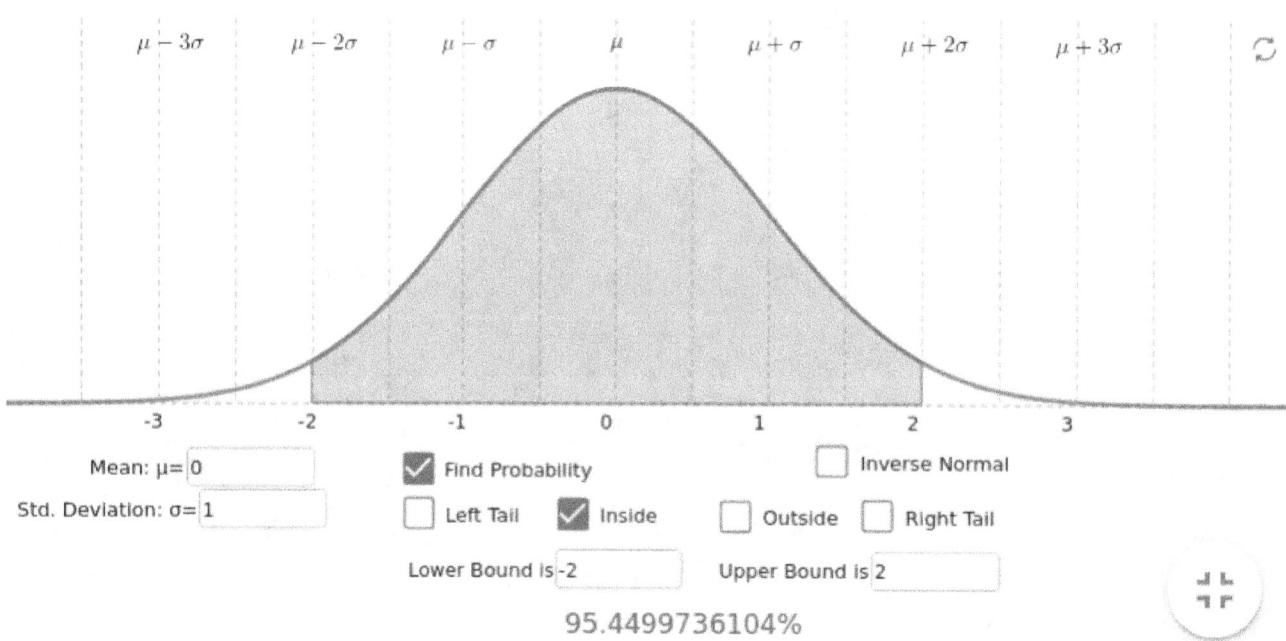

We see that approximately 95% of the data in a normal distribution is within two standard deviations of the mean.

Example: The length of a human pregnancy is normally distributed with a mean of 272 days with a standard deviation of 9 days.

a) State the random variable

b) Find the probability of a pregnancy lasting more than 280 days.

c) Find the probability of a pregnancy lasting less than 250 days.

d) Find the probability that a pregnancy lasts between 265 and 280 days.

e) Find the length of pregnancy that 10% of all pregnancies last less than.

f) Suppose you meet a woman who says that she was pregnant for less than 250 days. Would this be unusual?

g) Find the length of pregnancy that the middle 40% of all pregnancies last within.

h) Find the length of pregnancy for the longest 15% of all pregnancies.

Example: SAT: $\mu = 514$, $\sigma = 117$, normally distributed

a) State the random variable.

b) Find the probability that a person has a SAT score over 700.

c) Find the probability that a person has a SAT score of less than 400.

d) Find the probability that a person has a SAT score between 500 and 650.

e) Find the SAT score that represents the top 1% of all scores.

f) Find the SAT scores that represent the middle 25% of all scores.

g) Find the SAT score that represents the bottom 30% of all scores.

## Section 6.4: Assessing Normality

1. What is the source of ( ) in the population? If it is lots of ( ) factors, each ( ) or ( ), then the distribution is probably normal.

2. Histogram: should roughly be ( ).

3. Outliers: should not be more than ( ). Use the 1.5×IQR rule.

4. Normal Probability Plot: if the points lie close to a ( ), then the data is approximately normal.

Example: Kiama Blowhole

Is the time (in Seconds) between Kiama Blowhole eruptions normally distributed?

| 83 | 51 | 87 | 60 | 28 | 95 | 8 | 27 |
|----|----|----|----|----|----|----|----|
| 15 | 10 | 18 | 16 | 29 | 54 | 91 | 8 |
| 17 | 55 | 10 | 35 | 47 | 77 | 36 | 17 |
| 21 | 36 | 18 | 40 | 10 | 7 | 34 | 27 |
| 28 | 56 | 8 | 25 | 68 | 146 | 89 | 18 |
| 73 | 69 | 9 | 37 | 10 | 82 | 29 | 8 |
| 60 | 61 | 61 | 18 | 169 | 25 | 8 | 26 |
| 11 | 83 | 11 | 42 | 17 | 14 | 9 | 12 |

| Histogram | Boxplot | Normal Quantile Plot |
|-----------|---------|----------------------|
|           |         |                      |

Conclusion:

Example: IQ scores

| 78 | 92 | 96 | 100 | 67 | 105 | 109 | 75 | 127 | 111 |
|---|---|---|---|---|---|---|---|---|---|
| 93 | 114 | 82 | 100 | 125 | 67 | 94 | 74 | 81 | 98 |
| 102 | 108 | 81 | 96 | 103 | 91 | 90 | 96 | 86 | 92 |
| 84 | 92 | 90 | 103 | 115 | 93 | 85 | 116 | 87 | 106 |
| 85 | 88 | 106 | 104 | 102 | 98 | 116 | 107 | 102 | 89 |

| Histogram | Boxplot | Normal Quantile Plot |
|---|---|---|
|  |  |  |

Conclusion:

**Section 6.5: Sampling Distribution and the Central Limit Theorem**

• **Statistical Inference:** to make accurate decisions about parameters from statistics.

• **Sampling Distribution:** how a sample statistic is distributed when repeated trials of size $n$ are taken.

## Sampling Distribution

• Suppose you have a random variable that has a population mean, $\mu$, and a population standard deviation, $\sigma$. If a sample of size $n$ is taken, then the sample mean $\bar{x}$, has a mean of $\mu_{\bar{x}} = \mu$ and standard deviation of $\sigma_{\bar{x}} = \dfrac{\sigma}{\sqrt{n}}$.

## Central Limit Theorem

• If the random variable has a normal distribution, $\bar{x}$ will also be normally distributed.
• If the random variable has any distribution, the distribution of $\bar{x}$ will become normally distributed as $n$ increases.
• How big does $n$ have to be? It depends on the shape of the original distribution, but **around 30** the sampling distribution usually gets pretty close to normal.

# Chapter 7: One-Sample Inference

## Two forms of Inference

**Hypothesis Testing:** Making a decision about a ◯_____◯ (populations) based on a ◯_____◯ (sample). We might ask if the statement is true or not. However, because we are dealing with samples our statements are about **random experiments**. So, instead we ask how ◯_____◯ the statement is.

**Confidence Interval:** Estimating a parameter (populations) based on a statistic (sample). We might ask for the value of a parameter. However, because we are dealing with samples our statements are about **random experiments**. So, instead we ask how ◯_____◯ is a ◯_____◯ of values?

## Section 7.1: Basics of Hypothesis Testing

- A hypothesis is an educated ◯_____◯ about a parameter.

- The intuitive way: "Let's find evidence to support our hypothesis."

- The way that really works: "Assume our idea is ◯_____◯ true. Gather evidence. If that evidence would be very ◯_____◯, assuming our idea is not true, then it is more likely our idea is ◯_____◯."

## Life of an XJ35 Battery

- The manufacturer of the XJ35 battery claims the mean life of the battery is 500 days with a standard deviation of 25 days.

- We don't think they last that long.

- If we can support our view, maybe we can get out of that contract.


- Random variable: Let $x$ be the life of a XJ35 battery.

- Parameter: Let $\mu$ be the mean life of all XJ35 batteries.

- Sample: pick some batteries. Avoid bias: use random sample.

- How many? Central limit theorem makes 30 a good starting size.

## Data: life of battery in days

| 491 | 485 | 503 | 492 | 482 | 490 |
|-----|-----|-----|-----|-----|-----|
| 489 | 495 | 497 | 487 | 493 | 480 |
| 482 | 504 | 501 | 486 | 478 | 492 |
| 482 | 502 | 485 | 503 | 497 | 500 |
| 488 | 475 | 478 | 490 | 487 | 486 |

- Some are longer, some are shorter.
- $\bar{x} = 490$ is promising, but could this just be a coincidence. Maybe we just happened to pick a lot of short-life batteries.

## Two Hypotheses

**Null Hypothesis**: ⬭ value, ⬭, or ⬭. The symbol used is $H_0$.

**Alternate Hypothesis**: what you want to ⬭. This is what you want to ⬭ as true if you ⬭ the null hypothesis. There are two symbols that are commonly used for the alternative hypothesis: $H_a$ or $H_1$. We will use $H_a$.

Setting up formal expressions

$H_0: \mu = \mu_0$

$H_a: \mu < \mu_0$  or  $H_a: \mu \neq \mu_0$  or  $H_a: \mu > \mu_0$

$\mu_0$ represents the ⬭ that the ⬭ says the ⬭ mean is actually equal to.

A hypothesis is just a ⬭ with a name.

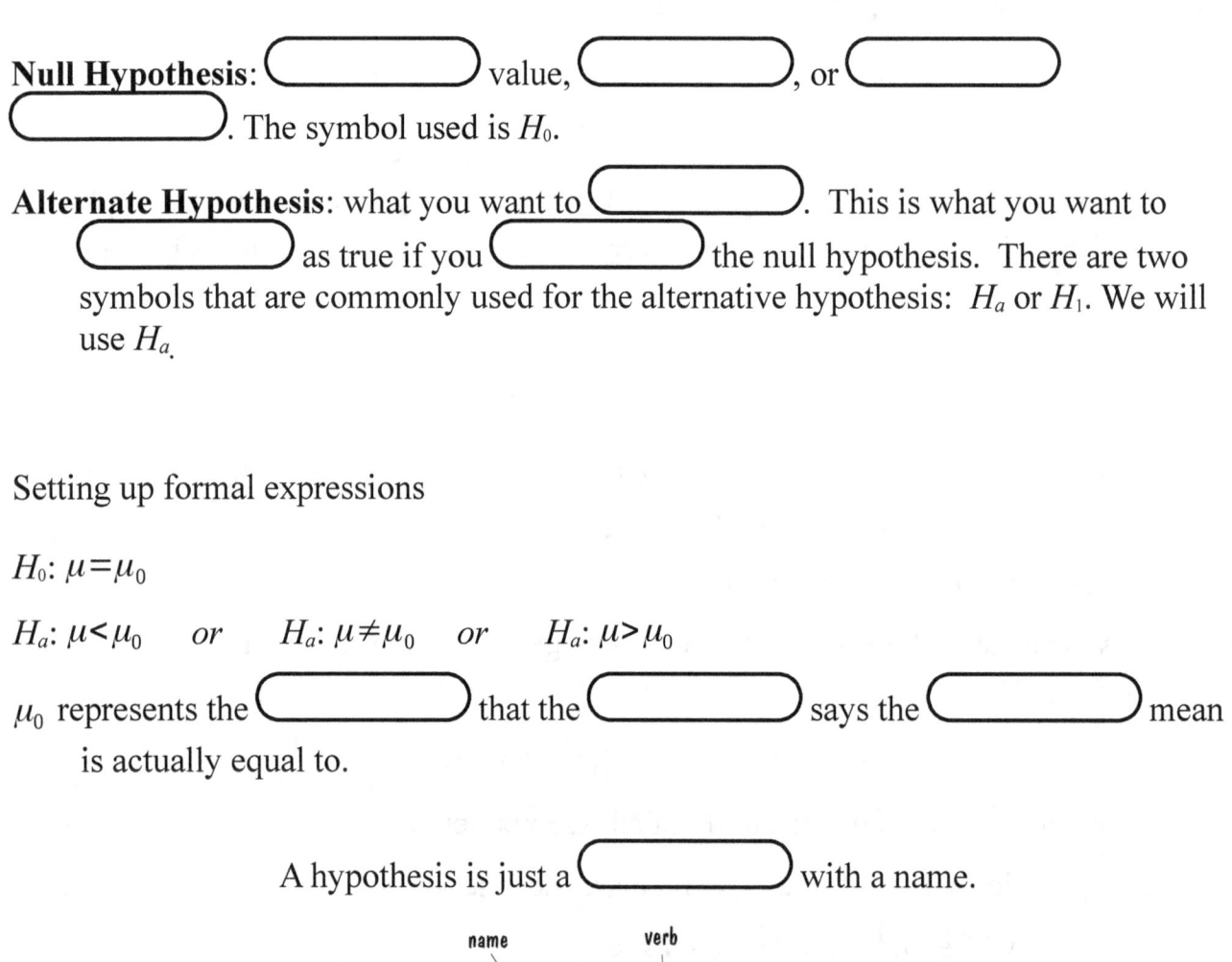

name → $H_0:$    verb → $\mu \stackrel{=}{=} \mu_0$

subject ↗    object ↖

Is the average less than 500?

- If our sample mean was 235, it would be a "slam dunk".
- If our sample mean was 435, "yeah, probably".
- If our sample mean was 483, or 498, "Gee, that might have been just chance."
- Where is the boundary? How do we find it?

Sampling distribution to the rescue.

If the mean really was 500, what different sample means would we get if we did repeated random samples of 30?

Sometimes we'd get more, sometimes less.

Distribution: **shape, center, spread**.

The **shape** of our distribution would be **normal**.

The **center** (mean) of our samples would be $\mu_{\bar{x}} = 500$.

The **spread** (standard deviation) of our samples would be $\sigma_{\bar{x}} = \dfrac{25}{\sqrt{30}} \approx 4.56$.

Distribution of sample means

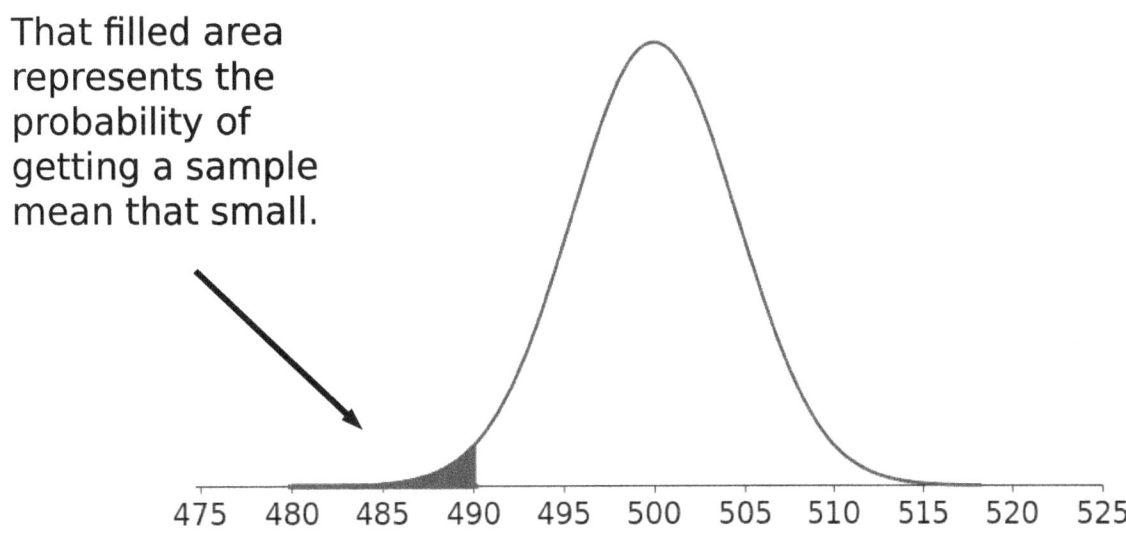

That filled area represents the probability of getting a sample mean that small.

475   480   485   490   495   500   505   510   515   520   525

## Return of the Z-Score

$$z = \frac{\bar{x} - \mu_0}{\sigma/\sqrt{n}} = \frac{490 - 500}{25/\sqrt{30}} \approx -2.19$$

Because that is more than 2 standard deviations it is unusual.

The probability is about 1.42%.

## Test Statistics & P-Value

**Test statistic:** $z = \frac{\bar{x} - \mu_0}{\sigma/\sqrt{n}}$ is calculated as part of testing of the hypothesis.

**p – value:** $\bigcirc$ that the $\bigcirc$ $\bigcirc$ will take on more extreme values than the $\bigcirc$ statistic, given that the null hypothesis is $\bigcirc$.

## Errors

|  | $H_0$ true | $H_0$ false |
|---|---|---|
| Reject | Type I error | No error |
| Fail to reject | No error | Type II error |

**Type I Error** is $\bigcirc$ $H_0$ when $H_0$ is $\bigcirc$, and
**Type II Error** is failing to $\bigcirc$ $H_0$ when $H_0$ is $\bigcirc$.

## Probabilities of errors

$\alpha = $ P(type I error) = P(reject $H_0 \mid H_0$)

> $\alpha$ is called the *level of significance*

> We can design our test to use $\alpha$ as our threshold.

$\beta = $ P(type II error) = P(fail to reject $H_0 \mid \sim H_0$)

*Power* $= 1 - \beta$.

For a certain sample size, $n$, if $\alpha$ increases, $\beta$ ⟨_____⟩.

For a certain level of significance, $\alpha$, if $n$ increases, $\beta$ ⟨_____⟩.

## Finding $\alpha$ and $\beta$

Now how do you find $\alpha$ and $\beta$? Generally, $\alpha$ is chosen. There are only three values that are usually picked for $\alpha$: 0.01, 0.05, and 0.10.

$\beta$ is very difficult to find, so usually it is not found. If you want to make sure it is small you take a sample as ⟨_____⟩ as you can afford, provided it is a ⟨_____⟩ sample. This is one use of the *Power*. You want $\beta$ to be small and then the *Power of the test* is ⟨_____⟩.

## Level of Evidence depends on the crime

If a type I error is really bad, then pick = 0.01.

If a type II error is really bad, then pick = 0.10

If neither error is bad, or both are equally bad, then pick = 0.05

The main thing is to always pick the $\alpha$ ⟨_____⟩ you collect the data and ⟨_____⟩ the test.

## Batteries revisited

1. State the random variable and the parameter in words

   Let $x$ be the life of a XJ35 battery.

   Let $\mu$ be the mean life of all XJ35 batteries.

2. State the null and alternative hypothesis and the level of significance

   $H_0$: $\mu = 500$

   $H_a$: $\mu < 500$

   $\alpha = 0.10$ (from above discussion about consequences)

3. State and check the assumptions for a hypothesis test

Every hypothesis has some assumptions that be met to make sure that the results of the hypothesis are valid. The assumptions are different for each test. This test has the following assumptions:

a.       A random sample of size $n$ is taken. This occurred in this example, since it was stated that a random sample of 30 battery lives were taken.

b.       The population standard deviation $\sigma$ is known. This is true, since it was given in the problem.

c.       The sample size is at least 30 or the population of the random variable is normally distributed. The sample size was 30, so this condition is met.

4. Find the sample statistic, test statistic, and p-value

a.       Sample statistic: $\bar{x} = 490$

b.       Test Statistic: $z = \dfrac{\bar{x} - \mu_0}{\sigma/\sqrt{n}} = \dfrac{490 - 500}{25/\sqrt{30}} \approx -2.19$

c.       p-value: (from "Z Test of Mean") area to the left of 490 on a normal curve with mean ($\mu$) of 500 and standard deviation of $\sigma_{\bar{x}} = \sigma/\sqrt{n} = 25/\sqrt{30} \approx 4.56$.

   $P(\mu = 490 | H_0) = 0.0142$

## 5. Conclusion

p-value $< \alpha$ since $0.0142 < 0.10$.

Reject $H_0$. The battery life probably was less than 500 days.

In general:

Reject $H_0$ if the p-value $< \alpha$

Fail to reject $H_0$ if the p-value $\geq \alpha$

## 6. Interpretation

Reject $H_0$, tentatively accept $H_a$

Tentatively state that the batteries last less than 500 days.

This is *statistically significant.*

### Process summary

1. State random variable and parameter in $\bigcirc\!\!\!\!\_\_\_\_\_\!\!\!\!\bigcirc$.

2. State null and alternate $\bigcirc\!\!\!\!\_\_\_\_\_\!\!\!\!\bigcirc$ and level of $\bigcirc\!\!\!\!\_\_\_\_\_\!\!\!\!\bigcirc$.

3. State and check $\bigcirc\!\!\!\!\_\_\_\_\_\!\!\!\!\bigcirc$ of the hypothesis test.

4. Find the $\bigcirc\!\!\!\!\_\_\_\_\_\!\!\!\!\bigcirc$ statistic, $\bigcirc\!\!\!\!\_\_\_\_\_\!\!\!\!\bigcirc$ statistic, and $\bigcirc\!\!\!\!\_\_\!\!\!\!\bigcirc$-value.

5. Make your conclusion: either reject $H_0$ (in favor of $H_a$) or fail to reject $H_0$ (not enough evidence for $H_a$). Never accept $H_0$!

## 7.2: One sample proportion test

To test a population proportion, there are a few things that need to be defined first. Usually, Greek letters are used for parameters and Latin letters for statistics. When talking about proportions, it makes sense to use $p$ for proportion. The Greek letter for $p$ is $\pi$, but that is too confusing to use. Instead, it is best to use $p$ for the population proportion. That means that a different symbol is needed for the sample proportion. The convention is to use, $\hat{p}$, known as p-hat. This way you know that $p$ is the population proportion, and that $\hat{p}$ is the sample proportion related to it.

1. State the random variable and the parameter in words.

   $x$ = number of successes

   $p$ = proportion of successes

2. State the null and alternative hypotheses and the level of significance.

   $H_0$: $p = p_0$, where $p_0$ is the *hypothetical proportion (assumed proportion)*

   $H_a$: $p < p_0$    *or*    $H_a$: $p \neq p_0$    *or*    $H_a$: $p > p_0$

   $\alpha$ is appropriate for the problem

3. State and check the assumptions for a hypothesis test

   a. An s.r.s. (simple random sample) of sample size $n$ is taken.

   b. The conditions for the binomial distribution are satisfied (two outcomes, fixed number of trials, independent)

   c. To determine the sampling distribution of $\hat{p}$, you need to show that $n\,p \geq 5$ and $n\,q \geq 5$, where $q = 1 - p$. If this requirement is met, then the sampling distribution of $\hat{p}$ is well approximated by a normal curve.

4. Find the sample statistic, test statistic, and p-value

   Sample Proportion: $\hat{p} = \dfrac{x}{n} = \dfrac{\text{number of successes}}{\text{number of trials}}$

   Test Statistic: $z = \dfrac{\hat{p} - p}{\sqrt{\dfrac{pq}{n}}}$

   p-value: We will use "Z Test of a Proportion"

5. **Conclusion** if p-value $< \alpha$, reject $H_0$ in favor of $H_a$, otherwise fail to reject $H_0$.

6. **Interpretation** This is where you interpret your result. The conclusion for a hypothesis test is that you either have enough evidence to show $H_a$ is true, or you do not have enough evidence to show $H_a$ is true.

## Z Test of a Proportion

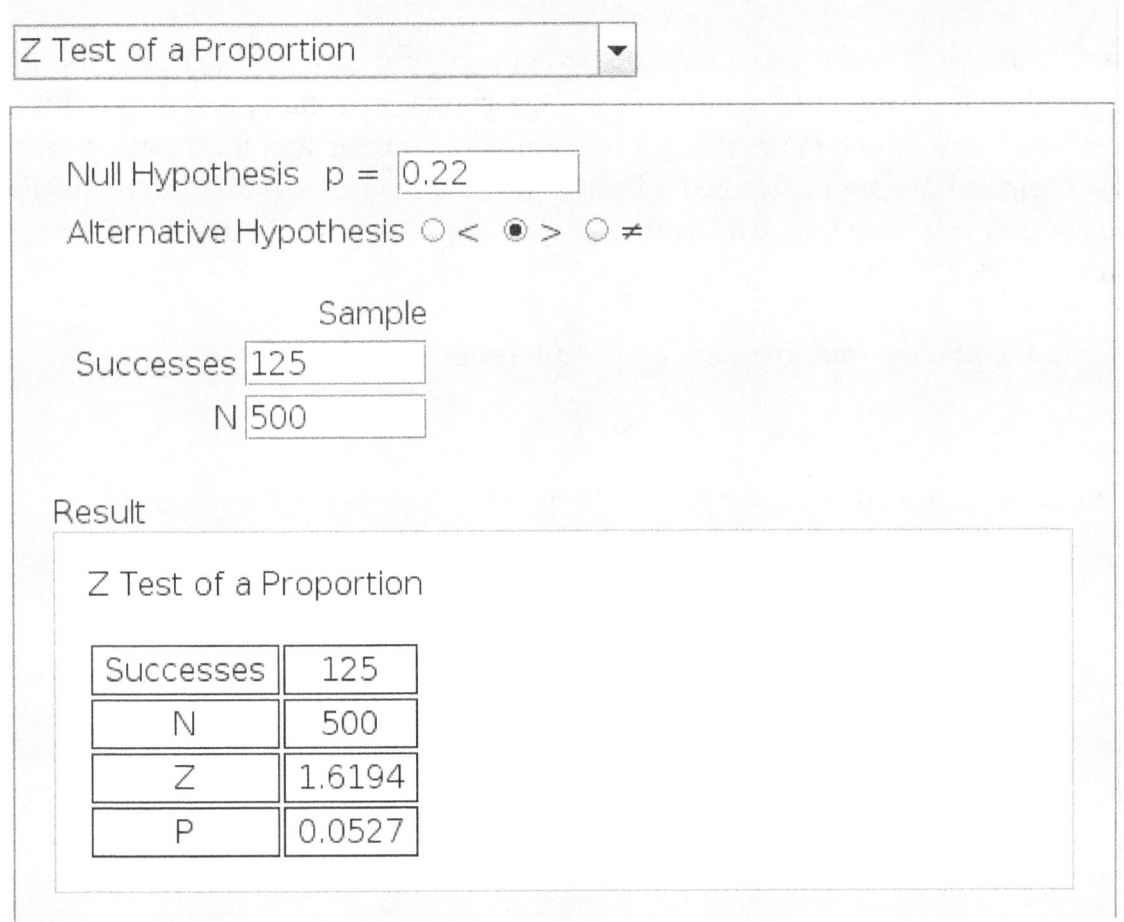

## Example

A concern was raised in Australia that the percentage of deaths of Aboriginal prisoners was higher than the percent of deaths of non-Aboriginal prisoners, which is 0.27%. A sample of six years (1990-1995) of data was collected, and it was found that out of 14,495 Aboriginal prisoners, 51 died ("Indigenous deaths in," 1996). Do the data provide enough evidence to show that the proportion of deaths of Aboriginal prisoners is more than 0.27%?

1.      State the random variable and the parameter in words.


2.      State the null and alternative hypotheses and the level of significance.


3.      State and check the assumptions for a hypothesis test




4. Find the sample statistic, test statistic, and p-value

    Sample:

    Test Statistic:

    p-value: Use "Z Test of a Proportion"



5.      Conclusion


6.      Interpretation

## Section 7.3 One-Sample Test for the Mean

It is time to go back to look at the test for the mean that was introduced in section 7.1 called the z-test. In the example, you knew the ⬭ standard deviation, $\sigma$. What if you do not know $\sigma$?

You could just use the sample standard deviation, $s$, as an ⬭ of $\sigma$. That means the test statistic is now: $\dfrac{\bar{x} - \mu_0}{s/\sqrt{n}}$.

Sources of variation

When we used $z = \dfrac{\bar{x} - \mu_0}{\sigma/\sqrt{n}}$, the only thing that varied from sample to sample was $\bar{x}$, the sample mean.

When we use $t = \dfrac{\bar{x} - \mu}{s/\sqrt{n}}$, two things can vary: $\bar{x}$, the ⬭ mean, and $s$, the ⬭ standard deviation.

$t$ follows a different (not normal) distribution, that is still *Bell shaped* and symmetric. It is actually a family of distributions, depending on the sample size.

Gosset Student's t-distribution

Theorem: t-distribution conditions

a.    A ⬭ sample of size $n$ is taken.

b.    The distribution of the random variable is normal or the sample size is over 30.

Then the distribution $t = \dfrac{\bar{x} - \mu}{s/\sqrt{n}}$ is a Gosset Student's t-distribution with $n - 1$ degrees of freedom.

# Why degree of freedom = $n - 1$

*If you five people and five chairs, the first four people have a choice of where they are sitting, but the last person does not. The last person has no freedom of where to sit. Only 5-1 = 4 people have freedom of choice.*

## Shapes of t-distributions

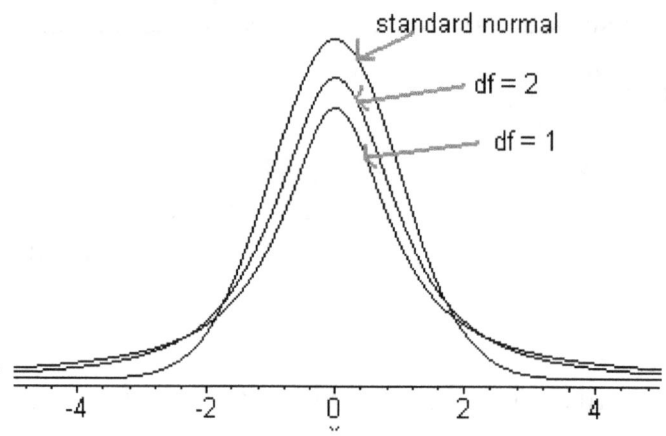

1. State the random variable and the parameter in words.

   $x$ is the random variable and $\mu$ is the mean of the random variable.

2. State the null and alternative hypotheses and the level of significance.

   $H_0$: $\mu = \mu_0$, where $\mu_0$ is the *hypothetical mean* (assumed mean).

   $H_a$: $\mu < \mu_0$   or   $H_a$: $\mu \neq \mu_0$   or   $H_a$: $\mu > \mu_0$

   $\alpha$ is appropriate for the problem

3. State and check the assumptions for a hypothesis test
   a.   A simple random sample of sample size $n$ is taken.
   b.   The population of the random variable is normally distributed, though the t-test is fairly robust to the condition if the sample size is large. This means that if this condition is not met, but your sample size is quite large (over 30), then the results of the t-test are valid.
   c.   The population standard deviation, $\sigma$, is unknown.

4. Find the sample statistic, test statistic, and p-value

   $$t = \frac{\bar{x} - \mu_0}{S_x / \sqrt{n}}, \text{ with degrees of freedom, } df = n - 1$$

   Technology will calculate the p-value for you if you tell it the alternate hypothesis.

5. Conclusion: If p-value $< \alpha$, reject $H_0$ in favor of $H_a$, otherwise fail to reject.

6. Interpretation: This is where you interpret in real world terms the conclusion to the test. The conclusion for a hypothesis test is that you either have enough evidence to show $H_a$ is true, or you do not have enough evidence to show $H_a$ is true.

*Note: if the assumptions behind this test are not valid, then the conclusions you make from the test are not valid. If you do not have a random sample, that is your fault. Make sure the sample you take is as random as you can make it following sampling techniques from chapter 1. If the population of the random variable is not normal, then take a sample larger than 30. If you cannot afford to do that, or if it is not logistically possible, then you do different tests called non-parametric tests. There is an entire course on non-parametric tests, and they will not be discussed in this class.*

# IQs of famous people

From the website of IQ of Famous People ("IQ of famous," 2013) a random number generator was used to pick a sample of 20.

| 158 | 180 | 150 | 137 | 109 |
|-----|-----|-----|-----|-----|
| 225 | 122 | 138 | 145 | 180 |
| 118 | 118 | 126 | 140 | 165 |
| 150 | 170 | 105 | 154 | 118 |

Does the data provide evidence at the 5% level that the IQ of a famous person is higher than the average IQ of 100?

## Lifespans of European Males

In 2011, the average life expectancy for a woman in Europe was 79.8 years. The data in table #7.3.2 are the life expectancies for men in European countries in 2011 ("WHO life expectancy," 2013). Do the data indicate that men's life expectancy is less than women's? Test at the 1% level.

| 73 | 79 | 67 | 78 | 69 | 66 | 78 | 74 |
|----|----|----|----|----|----|----|----|
| 71 | 74 | 79 | 75 | 77 | 71 | 78 | 78 |
| 68 | 78 | 78 | 71 | 81 | 79 | 80 | 80 |
| 62 | 65 | 69 | 68 | 79 | 79 | 79 | 73 |
| 79 | 79 | 72 | 77 | 67 | 70 | 63 | 82 |
| 72 | 72 | 77 | 79 | 80 | 80 | 67 | 73 |
| 73 | 60 | 65 | 79 | 66 |    |    |    |

# Chapter 8: Estimation

## Confidence Intervals

### Sec. 8.1 Basics of Confidence Intervals

Point estimator: just the statistic that you have calculated previously.

    How accurate?

    What was the sample size?

    How likely is it to be right?

Confidence Interval is always two things:

    An interval: $\hat{\theta} \pm E$, where $\hat{\theta}$ is the point estimate, and $E$ is the margin of error.

    A level of confidence, $1-\alpha$

Interpretation of C. I.s

If you have a 95% confidence interval of $0.65 < p < 0.73$, then you would say, "there is a 95% ⬭ that the interval 0.65 to 0.73 contains the ⬭ ⬭⬭."

The probability is whether or not we have captured the population proportion in the confidence interval.

Ex 1: State as an interval and as a sentence: $\bar{x}=25$ and $E=8$ with $\alpha=0.01$.

### Affect of Confidence Level on Width

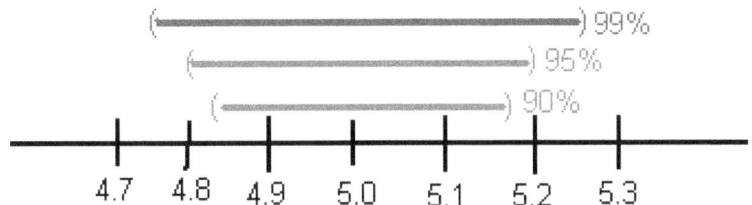

There's a trade-off between ⬭ and ⬭ level. You can be very ⬭ about your answer but your answer will not be very ⬭. Or you can have a ⬭ answer (small margin of error) but not be very ⬭ about your answer.

Ex 2: There is a 95% chance that_____.

### Affect of Sample Size on Width

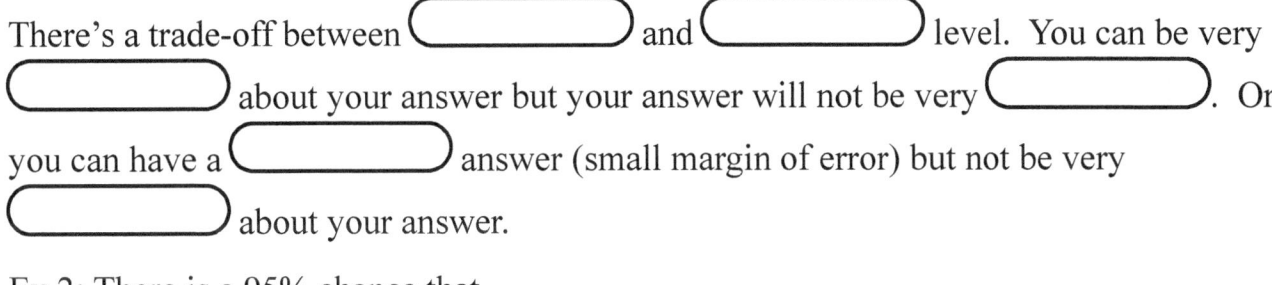

A larger sample size from a representative sample makes the width of the interval narrower. Large samples are closer to the true population so the point estimate is pretty close to the true value.

Ex 3: Suppose you compute a confidence interval with a sample size of 25. What will happen to the confidence interval if the sample size increases to 50?

## Sec. 8.2 One-Sample Interval for the Proportion

1) Random variable and parameter:

$x$ = number of successes.

$p$ = proportion of successes.

2) Assumptions:

a) SRS of size n

b) Conditions of binomial distribution are satisfied

c) $n\hat{p} \geq 5$ and $n\hat{q} \geq 5$ ($x \geq 5$ and $n-x \geq 5$)

3) Find the sample statistic and the confidence interval

$\hat{p} - E < p < \hat{p} + E$

$p$ = population proportion

$\hat{p}$ = sample proportion

$\hat{p} = \dfrac{x}{n}$

$n$ = number of sample values

$E$ = margin of error

$z_C$ = critical value where C = 1 - $\alpha$

$\hat{q} = 1 - \hat{p}$

$E = z_C \sqrt{\dfrac{\hat{p}\hat{q}}{n}}$

4)     Statistical Interpretation: In general this looks like: "there is a C% chance that $\hat{p} - E < p < \hat{p} + E$ contains the true proportion."

5)     Real World Interpretation: This is where you state what interval contains the true proportion.

# Critical Value ($z_C$)

The critical value is a value from the standard normal distribution. Since a
$\bigcirc$ $\bigcirc$ is found by adding and subtracting a $\bigcirc$ of
$\bigcirc$ from the $\bigcirc$ $\bigcirc$, and the interval has a
$\bigcirc$ of containing the $\bigcirc$ proportion, then you can think of this
as the statement $P(\hat{p}-E<p<\hat{p}+E)=C$. You can use the Normal Distribution
Calculator to find the critical value. Some become familiar (1.96 for 95%).

Ex 4: Use the Normal Distribution Calculator to find the critical value for $C=95\%$,
$C=90\%$, and $C=99\%$.

# Standard error

- One important part of the calculation was, $\sqrt{\dfrac{\hat{p}\hat{q}}{n}}$ which we refer to a "Standard
  Error". You may think of it as the standard deviation of the sampling distribution,
  although that may not be strictly true for technical reasons.

- The confidence interval is always a *Point Estimate* plus or minus a *Margin of Error*.

- The *Margin of Error* is always a critical value times a *Standard Error*.

Ex 5:

Use a random sample to estimate a population proportion $p$. Find the 90% confidence interval for a sample of size 500 with 125 successes.

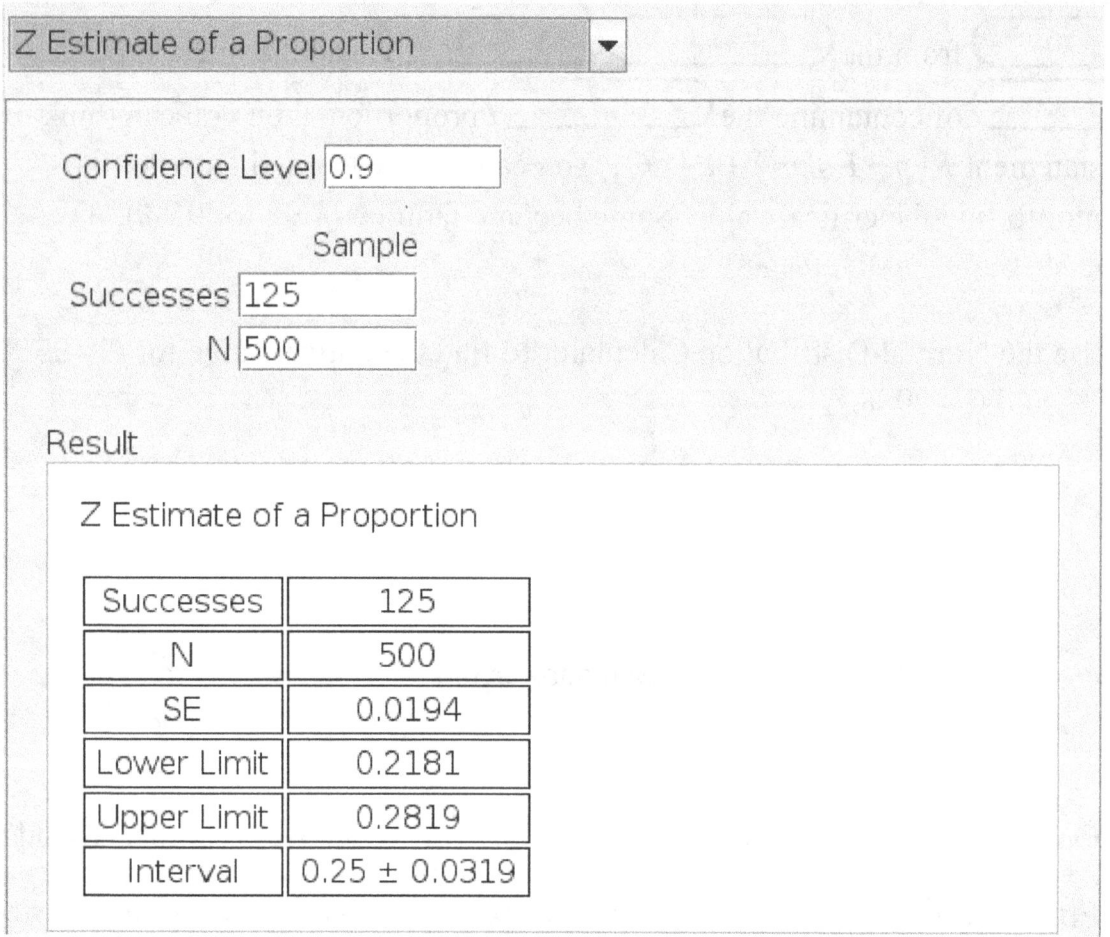

Note: $x$ must be entered as an integer. If you have a problem where a sample proportion, $\hat{p}$, is given, you can "reconstruct" the $x$ as $x = n\hat{p}$. But you must round to the closest integer.

Ex 6: Out of a random sample of 159 adults, 68% had kids. Based on this, construct a 95% confidence interval for the true population proportion of people with kids.

## Sec. 8.3 One-Sample Interval for the Mean

1) Random variable and parameter

   $x$ = random variable

   $\mu$ = mean of the random variable

2) Assumptions:

   a. A random sample of size n is taken.

   b. The population is normally distributed or the sample size is over 30.

3) Confidence interval

   $\bar{x} - E < \mu < \bar{x} + E$

   $$E = t_C \frac{s}{\sqrt{n}}$$

   $\bar{x}$ is the point estimator for $\mu$

   $s$ is the sample standard deviation

   $n$ is the sample size

   $E$ is the margin of error

   $t_C$ is the critical value where $C = 1 - \alpha$

4) Statistical Interpretation: In general this looks like, "there is a C% chance that the interval $\bar{x} - E < \mu < \bar{x} + E$ contains the true mean."

5) Real World Interpretation: This is where you state what interval contains the true mean.

Getting $t_C$:

- Critical t values are often retrieved from a table, as it was only recently that it became easy to do the inverse t calculations

- Use the Gosset Student T Distribution Calculator with the $df = n - 1$, provide the probability as an "inside" area, and you will get -t and +t

Ex 7: Find the critical value, $t_C$, when $C = 90\%$ for a random sample of 350.

Getting the Confidence Interval:

- If you provide the ⬭⬭ and use $s/\sqrt{n}$ for the standard deviation, the Gosset Student T Distribution Calculator will calculate the confidence interval for you. Again, you have a probability with the $df = n - 1$. Find the probability as an area "inside" and you will get the interval.

Ex 8: Use the Gosset Student T Distribution Calculator to find the confidence interval when $C = 95\%$ for a random sample of 29, $\bar{x} = 56$ and $S_x = 8.6$ from a normally distributed population.

- But there's an even easier way: T Estimate of a Mean

Ex 9: Use T Estimate of a Mean to find the confidence interval when $C = 95\%$ for a random sample of 29, $\bar{x} = 56$ and $S_x = 8.6$ from a normally distributed population.

Ex 10:

A random sample of 20 IQ scores of famous people was taken information from the website of IQ of Famous People and then using a random number generator to pick 20 of them. Find a 98% confidence interval for the IQ of a famous person.

| | | | | |
|-----|-----|-----|-----|-----|
| 158 | 180 | 150 | 137 | 109 |
| 225 | 122 | 138 | 145 | 180 |
| 118 | 118 | 126 | 140 | 165 |
| 150 | 170 | 105 | 154 | 118 |

Ex 10:

The data in table #8.3.3 are the life expectancies for men in European countries in 2011 ("WHO life expectancy," 2013). Find the 99% confident interval for the mean life expectancy of men in Europe. Use the data analysis.

| 73 | 79 | 67 | 78 | 69 | 66 | 78 | 74 |
|----|----|----|----|----|----|----|----|
| 71 | 74 | 79 | 75 | 77 | 71 | 78 | 78 |
| 68 | 78 | 78 | 71 | 81 | 79 | 80 | 80 |
| 62 | 65 | 69 | 68 | 79 | 79 | 79 | 73 |
| 79 | 79 | 72 | 77 | 67 | 70 | 63 | 82 |
| 72 | 72 | 77 | 79 | 80 | 80 | 67 | 73 |
| 73 | 60 | 65 | 79 | 66 |    |    |    |

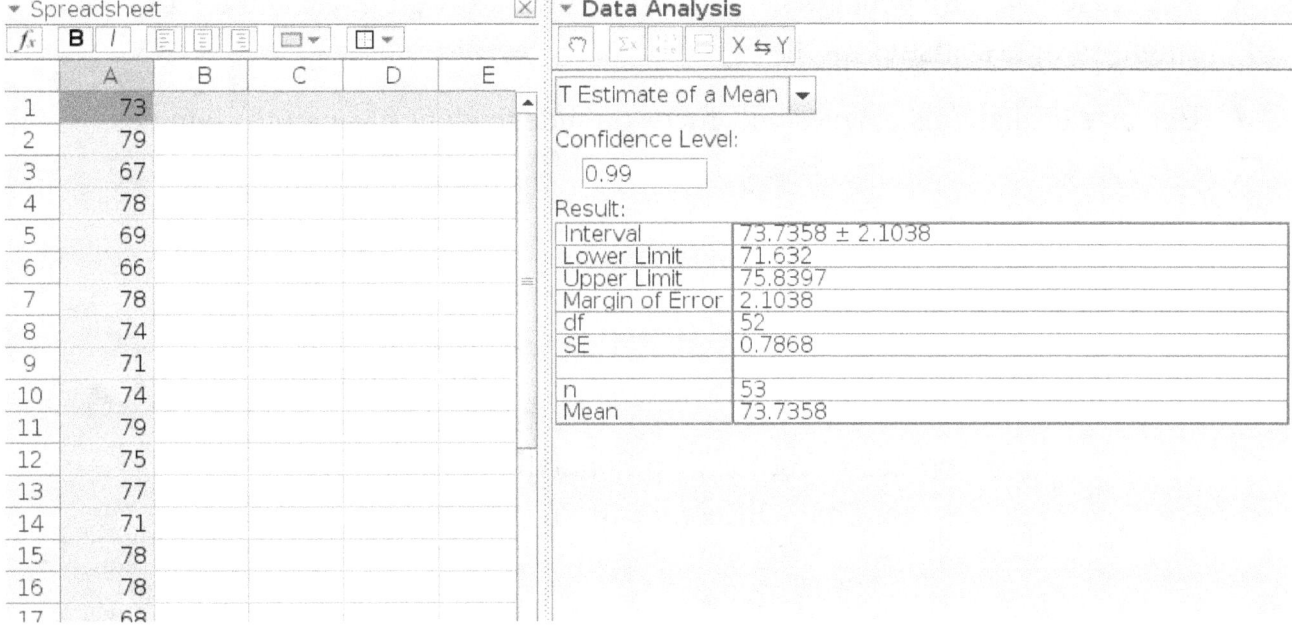

# Chapter 9

## Two-Sample Inference

Comparing: Does it work?  Does *this* affect *that*?

Is the data in pairs?

- In paired inference each measurement ⬭ with another measurement. This could come from a ⬭ and ⬭ study, a study of ⬭, or a study of the right and left arms.

- There are always ⬭ the same ⬭ of measurements in each data set – and each measurement is matched or paired with with just one measurement in the other list.

- The reason we care about pairing is this: if they are paired, we can treat them as single-sample statistics on the *differences*.

- Because of that, this chapter will focus on independent (un-paired) samples.

## Section 9.1 Two Proportions

Hypothesis Test for Two Population Proportions
Z Test, Difference of Proportions

1)  State random variables, $x_1$ and $x_2$, and parameters, $p_1$ and $p_2$, in words.

2)  Hypotheses and $\alpha$

$H_0$: $p_1 = p_2$        (Note that this is the same as: $p_1 - p_2 = 0$.)

$H_a$: $p_1 < p_2$   or   $p_1 \neq p_2$   or   $p_1 > p_2$

3)    Assumptions

   a) Simple random sample of both populations

   b) Samples of both populations are independent

   c) Binomial conditions: fixed sample size, two outcomes, independent

   d) Notice that the null hypothesis no longer provides a claim for the value of the population proportions. Since we are not testing a claim for $p_1$ and $p_2$, we must use $\hat{p}_1$ and $\hat{p}_2$ instead. Also, since $n_1 \hat{p}_1 = n_1 \dfrac{x_1}{n_1} = x_1$ we only need to shown that the number of success, $x_1$ and $x_2$, failures, $n_1 - x_1$ and $n_2 - x_2$, are more than 5.

4)    Find the sample statistic, test statistic, and p-value for the pooled proportion:

$$\bar{p} = \frac{x_1 + x_2}{n_1 + n_2}, \text{ and } \bar{q} = 1 - \bar{p}$$

The book shows:
$$z = \frac{(\hat{p}_1 - \hat{p}_2) - (p_1 - p_2)}{\sqrt{\dfrac{\bar{p}\bar{q}}{n_1} + \dfrac{\bar{p}\bar{q}}{n_2}}}$$

We use
$$z = \frac{\hat{p}_1 - \hat{p}_2}{\sqrt{\dfrac{\bar{p}\bar{q}}{n_1} + \dfrac{\bar{p}\bar{q}}{n_2}}} \text{ because we are assuming that } p_1 = p_2$$

p-value = area of left, right, or both tails, depending on the alternate hypothesis.

That is, the p-value is the probability of getting results as extreme as the sample, while assuming $H_0$.

5)    Conclusion.  As always, if p-value $< \alpha$, reject the null hypothesis in favor of the alternate.  There is sufficient evidence to support the alternate hypothesis. Otherwise, there is not enough evidence to support the alternate hypothesis at the stated level $\alpha$.

6)    Interpretation: what does this conclusion imply in the context of the problem?

# Confidence Interval
## Z Estimate, Difference of Proportions

Steps 1 and 3 are the same as for a 2-proportion hypothesis test.

C.I. = point estimate $\pm$ margin of error

point estimate = $\hat{p}_1 - \hat{p}_2$

margin of error = $z_C\sqrt{\dfrac{\hat{p}_1\hat{q}_1}{n_1} + \dfrac{\hat{p}_2\hat{q}_2}{n_2}}$ where $z_C$ comes from the table or the Normal Distribution Calculator.

Confidence interval estimate of $p_1 - p_2$ is $(\hat{p}_1 - \hat{p}_2) - E < p_1 - p_2 < (\hat{p}_1 - \hat{p}_2) + E$

## Example: Cheating Husbands

Do more husbands cheat on their wives more than wives cheat on the husbands ("Statistics brain," 2013)? Suppose you take a group of 1000 randomly selected husbands and find that 231 had cheated on their wives. Suppose in a group of 1200 randomly selected wives, 176 cheated on their husbands. Does the data show that the proportion of husbands who cheat on their wives is more than the proportion of wives who cheat on their husbands? Test at the 5% level.

Z Test, Difference of Proportions ▾

Null Hypothesis p₁ - p₂ = 0
Alternative Hypothesis ○ <  ⦿ >  ○ ≠

|  | Sample 1 | | Sample 2 |
| --- | --- | --- | --- |
| Successes | 231 | Successes | 176 |
| N | 1000 | N | 1200 |

Result

Z Test, Difference of Proportions

|  | Sample 1 | Sample 2 |
| --- | --- | --- |
| Successes | 231 | 176 |
| N | 1000 | 1200 |
| SE | 0.0166 | |
| Z | 5.0724 | |
| P | 0 | |

1)   State random variables and parameters in words.

2)   Hypotheses and $\alpha$

3)   Assumptions

  a)

  b)

  c)

  d)

4)   Find the sample statistic, test statistic, and p-value

5)   Conclusion:

6)   Interpretation:

Estimate difference in cheating rates using a 95% confidence interval.

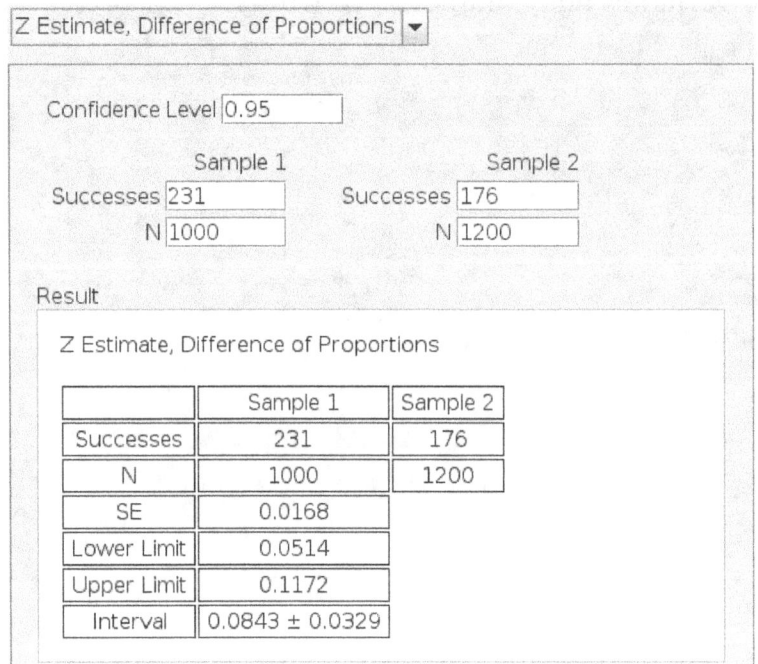

Steps 1 and 2 are the same as for the hypothesis test.

3)      Find the sample statistic and the confidence interval

4)      Statistical Interpretation:

5)      Real World Interpretation:

# Section 9.2 Paired Samples for Two Means

- Make sure you can differentiate between *matched* (*paired* or *dependent*) samples and *independent* samples.

- If the lists are of different lengths, they are *independent*.

## Shortcut for matched pairs

- Just create a new list of differences, and treat that list as a single-sample statistics problem (hypothesis or C.I., means or proportions)

- Enter the samples then form a new column of differences.

- Use the methods of Chapter 7 and 8.

## Hypothesis Test (T Test of a Mean)

The assumptions for the hypothesis test are:

a. A simple random sample of sample size $n$ is taken.

b. The population of the random variable is normally distributed, though the t-test is fairly robust to the condition if the sample size is large. This means that if this condition isn't met, but your sample size is quite large (over 30), then the results of the t-test are valid.

c. The population standard deviation, $\sigma$, is unknown.

The sample statistic, test statistic, and p-value are:

Difference: $d = x_1 - x_2$ for each pair

Sample mean of the differences: $\bar{d} = \dfrac{\sum d}{n}$

Standard deviation of the differences: $s_d = \sqrt{\dfrac{\sum (d - \bar{d})^2}{n-1}}$

Number of pairs: $n$

Test Statistic: $t = \dfrac{\bar{d} - \mu_d}{\dfrac{s_d}{\sqrt{n}}}$ with $df = n - 1$ and $\mu_d = 0$ in most cases.

P-value: Use T Test of a Mean

Ex: The cholesterol level of patients who had heart attacks was measured two days after the heart attack and then again four days after the heart attack. The researchers want to see if the cholesterol level of patients who have heart attacks reduces as the time since their heart attack increases. Do the data show that the mean cholesterol level of patients that have had a heart attack reduces as the time increases since their heart attack? Test at the 1% level.

| Patient | Cholesterol Level Day 2 | Cholesterol Level Day 4 | Patient | Cholesterol Level Day 2 | Cholesterol Level Day 4 |
|---|---|---|---|---|---|
| 1 | 270 | 218 | 16 | 282 | 294 |
| 2 | 236 | 234 | 17 | 234 | 220 |
| 3 | 210 | 214 | 18 | 224 | 200 |
| 4 | 142 | 116 | 19 | 276 | 220 |
| 5 | 280 | 200 | 20 | 282 | 186 |
| 6 | 272 | 276 | 21 | 360 | 352 |
| 7 | 160 | 146 | 22 | 310 | 202 |
| 8 | 220 | 182 | 23 | 280 | 218 |
| 9 | 226 | 238 | 24 | 278 | 248 |
| 10 | 242 | 288 | 25 | 288 | 278 |
| 11 | 186 | 190 | 26 | 288 | 248 |
| 12 | 266 | 236 | 27 | 244 | 270 |
| 13 | 206 | 244 | 28 | 236 | 242 |
| 14 | 318 | 258 | 29 | **228** | **206** |
| 15 | 294 | 240 | 30 | **235** | **221** |

(This is Table #9.2.4: Cholesterol Levels (in mg/dL) of Heart Attack Patients on page 307 with two additional patients added and shown bold.)

1.  State the random variable and the parameter in words.

2.  State the null and alternative hypotheses and the level of significance.

3.  State and check the assumptions for a hypothesis test

    a.

    b.

    c.

4.  Find the sample statistic, test statistic, and p-value

5.  Conclusion:

6.  Interpretation:

# Confidence Intervals (T Estimate of a Mean)

Steps 1 and 3 are the same as for a 2-proportion hypothesis test.

3. Find the sample statistic and confidence interval

The sample statistics are the same:

Difference: $d = x_1 - x_2$

Sample mean of the differences: $\bar{d} = \dfrac{\sum d}{n}$

Standard deviation of the differences: $s_d = \sqrt{\dfrac{\sum (d - \bar{d})^2}{n-1}}$

Number of pairs: $n$

The confidence interval estimate of the difference $\mu_d = \mu_1 - \mu_2$ is:

$$\bar{d} - E < \mu_d < \bar{d} + E, \quad E = t_C \frac{s_d}{\sqrt{n}} \text{ where } t_C \text{ is the critical value and the}$$

degrees of freedom is $df = n - 1$.

4. Statistical Interpretation
5. Real World Interpretation

Ex.: Use the data to estimate the true mean difference between cholesterol levels 2 days and 4 days after a heart attack at 95% confidence.

3. Find the sample statistic and confidence interval

4. Statistical Interpretation

5. Real World Interpretation

## Section 9.3 Independent Samples for Two Means

This section will look at how to analyze two *independent* samples. The only difference with the independent t-test is that there are actually two different formulas to use depending on an assumption about the population variances.

Hypothesis Test for Independent t-Test (T Test, Difference of Means)

1) Variables & parameters: $x_1, x_2, \mu_1, \mu_2$.

2) Hypotheses & $\alpha$

$$H_0: \mu_1 = \mu_2$$

$$H_a: \mu_1 < \mu_2 \quad \text{or} \quad \mu_1 \neq \mu_2 \quad \text{or} \quad \mu_1 > \mu_2$$

3) Assumptions:

    a) s.r.s. (the test is more reliable when the sizes are equal)

    b) The two samples are independent.

    c) Either the populations are known to be normally distributed or the sample size is greater than 30.

    d) If the population variances are known to be equal then pool the sample standard deviation. Otherwise, do not pool the sample standard deviations.

4) Sample statistic, test statistic, degrees of freedom and p-value, assuming $\mu_1 = \mu_2$:

$$t = \frac{\bar{x}_1 - \bar{x}_2}{\sqrt{\dfrac{s_1^2}{n_1} + \dfrac{s_2^2}{n_2}}}$$

    (*df* is found with a formula or table.)

p-value = area of left, right, or both tails, depending on the alternate hypothesis.

That is, the p-value is the probability of getting results as extreme as the sample, assuming $H_0$.

5) Conclusion. As always, if p-value $< \alpha$, reject the null hypothesis in favor of the alternate. There is sufficient evidence to support the alternate hypothesis. Otherwise, there is not enough evidence to support the alternate hypothesis at the stated level $\alpha$.

6) Interpretation: what does this conclusion imply in the context of the problem?

Ex.: The cholesterol level (mg/dL) of patients who had heart attacks was measured two days after the heart attack. The researchers want to see if patients who have heart attacks have higher cholesterol levels than healthy people, so they also measured the cholesterol level of healthy adults who show no signs of heart disease. Does the data show that people who have had heart attacks have higher cholesterol levels over patients that have not had heart attacks? Test at the 1% level. (Note that the "Cholesterol Level of Heart Attack Patients" is the same data that we used previously as "Cholesterol Level Day 2." This is Table #9.3.1: Cholesterol Levels in mg/dL on page 316 with four additional samples added and shown bold.)

| Cholesterol Level of Heart Attack Patients | Cholesterol Level of Healthy Individual | Cholesterol Level of Heart Attack Patients | Cholesterol Level of Healthy Individual | Cholesterol Level of Heart Attack Patients | Cholesterol Level of Healthy Individual |
|---|---|---|---|---|---|
| 270 | 196 | 266 | 198 | 280 | 230 |
| 236 | 232 | 206 | 182 | 278 | 186 |
| 210 | 200 | 318 | 238 | 288 | 162 |
| 142 | 242 | 294 | 198 | 288 | 182 |
| 280 | 206 | 282 | 188 | 244 | 218 |
| 272 | 178 | 234 | 166 | 236 | 170 |
| 160 | 184 | 224 | 204 | **228** | 200 |
| 220 | 198 | 276 | 182 | **235** | 176 |
| 226 | 160 | 282 | 178 | | **204** |
| 242 | 182 | 360 | 212 | | **283** |
| 186 | 182 | 310 | 164 | | |

T Test, Difference of Means ▼
☐ Pooled

Null Hypothesis: $\mu_1 - \mu_2 = 0$

Alternative Hypothesis: $\mu_1 - \mu_2 > 0$ ▼

| | | Mean | s | n |
|---|---|---|---|---|
| Sample 1 | Column A ▼ | 252.4333 | 46.3954 | 30 |
| Sample 2 | Column B ▼ | 196.2813 | 26.8199 | 32 |

| Difference | P | t | SE | df |
|---|---|---|---|---|
| 56.1521 | 0 | 5.7846 | 9.7072 | 45.8105 |

Note: We did not pool the standard deviation since we are not assuming the population variances are equal.

1) Variables & parameters:




2) Hypotheses & $\alpha$:




3) Assumptions:

    a)

    b)

    c)

    d)


4) Sample statistic, test statistic, degrees of freedom and p-value:






5) Conclusion:




6) Interpretation:

**Confidence Interval for $\mu_1$ - $\mu_2$ (T Estimate, Difference of Means)**

Steps 1 and 3 are the same as for a 2-sample hypothesis test.

C.I. = point estimate $\pm$ margin of error

point estimate = $\bar{x}_1 - \bar{x}_2$

margin of error = $t_C \sqrt{\dfrac{s_1^2}{n_1} + \dfrac{s_2^2}{n_2}}$ where $t_C$

Confidence interval estimate of $\mu_1 - \mu_2$ is $\left(\bar{x}_1 - \bar{x}_2\right) - E < \mu_1 - \mu_2 < \left(\bar{x}_1 - \bar{x}_2\right) + E$

Ex.: Use the same data as previous example. Find a 99% confidence interval for the mean difference in cholesterol levels between heart attack patients and healthy individuals.

T Estimate, Difference of Means ▼

☐ Pooled

Confidence Level: 0.99

|  |  | Mean | s | n |
|---|---|---|---|---|
| Sample 1 | Column A ▼ | 252.4333 | 46.3954 | 30 |
| Sample 2 | Column B ▼ | 196.2813 | 26.8199 | 32 |

| Difference | ME | Lower Limit | Upper Limit | SE | df |
|---|---|---|---|---|---|
| 56.1521 | 26.088 | 30.0641 | 82.24 | 9.7072 | 45.8105 |

Steps 1 and 3 are the same as for a 2-sample hypothesis test.

3)      Find the sample statistic and the confidence interval

4)      Statistical Interpretation:

5)      Real World Interpretation:

So far we have been assuming that the two population variances are not equal, $\sigma_1^2 \neq \sigma_2^2$. That is why we selected "No" when asked to pool the sample means. If the population variances were equal, we could have used a simpler method to find the t-score and margin of error. However, since we only know the sample variations and not the population variations, if we wish to use the simpler method, we must first show that the populations variances are equal, $\sigma_1^2 = \sigma_2^2$. We use an F-test to do this. To show that the variances are equal, we must show that the ratio of the sample variances is not unusual (probability is greater than 0.05). In other words, make sure the following is true: $P(F > \sigma_1^2 / \sigma_2^2) \geq 0.05$ (or $P(F > \sigma_2^2 / \sigma_1^2) \geq 0.05$ so that the larger variance is in the numerator). To find the probability use the F Cumulative Distribution Function Calculator. If the probability is greater than equal to 0.05, with the larger variance in the numerator, then you may use these equations for the t-score and margin of error.

That said, you do not need to use the simpler tests. Beginning on page 324, our text shows you how to use the F-test and the simpler equations. We would first do the F-test and if the probability is greater than 0.05 we then select "Pooled" in "T Test, Difference of Means" or "T Estimate, Difference of Means."

| The pooled standard deviation is $s_p = \sqrt{\dfrac{(n_1 - 1)s_1^2 + (n_2 - 1)s_2^2}{(n_1 - 1) + (n_2 - 1)}}$. | |
|---|---|
| **Independent Samples for Two Means** $t = \dfrac{\bar{x}_1 - \bar{x}_2}{s_p \sqrt{\dfrac{1}{n_1} + \dfrac{1}{n_2}}}$ where $df = n_1 + n_2 - 2$. <br><br> p-value is found using: <br><br> T Test, Difference of Means | **Confidence Interval for $\mu_1$ - $\mu_2$** $E = t_C s_p \sqrt{\dfrac{1}{n_1} + \dfrac{1}{n_2}}$, where $t_C$ is found using Gosset Student T Distribution Calculator or tables. <br><br> Confidence interval estimate of $\mu_1 - \mu_2$ is $(\bar{x}_1 - \bar{x}_2) - E < \mu_1 - \mu_2 < (\bar{x}_1 - \bar{x}_2) + E$ <br><br> Or using: T Estimate, Difference of Means |

# Chapter 10

## Regression and Correlation

### Two Quantitative Variables

- Is there a relationship between two variables?

- What is the shape of the relationship?

- How strong is that relationship?

Beer: calories and alcohol

| Brand | Alcohol Content | Calories in 12 oz |
|---|---|---|
| Big Sky Scape Goat Pale Ale | 4.70% | 163 |
| Sierra Nevada Harvest Ale | 6.70% | 215 |
| Steel Reserve | 8.10% | 222 |
| O'Doul's | 0.40% | 70 |
| Coors Light | 4.15% | 104 |
| Genesee Cream Ale | 5.10% | 162 |
| Sierra Nevada Summerfest Beer | 5.00% | 158 |
| Michelob Beer | 5.00% | 155 |
| Flying Dog Doggie Style | 4.70% | 158 |
| Big Sky I.P.A. | 6.20% | 195 |

Put the independent data in column A and the dependent data in column B. Select both columns and choose "Two Variable Regression Analysis."

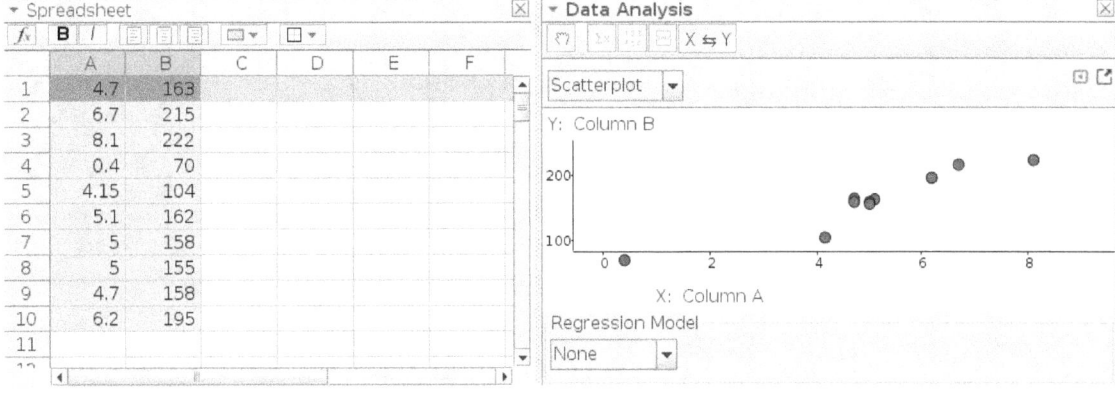

Scatterplot

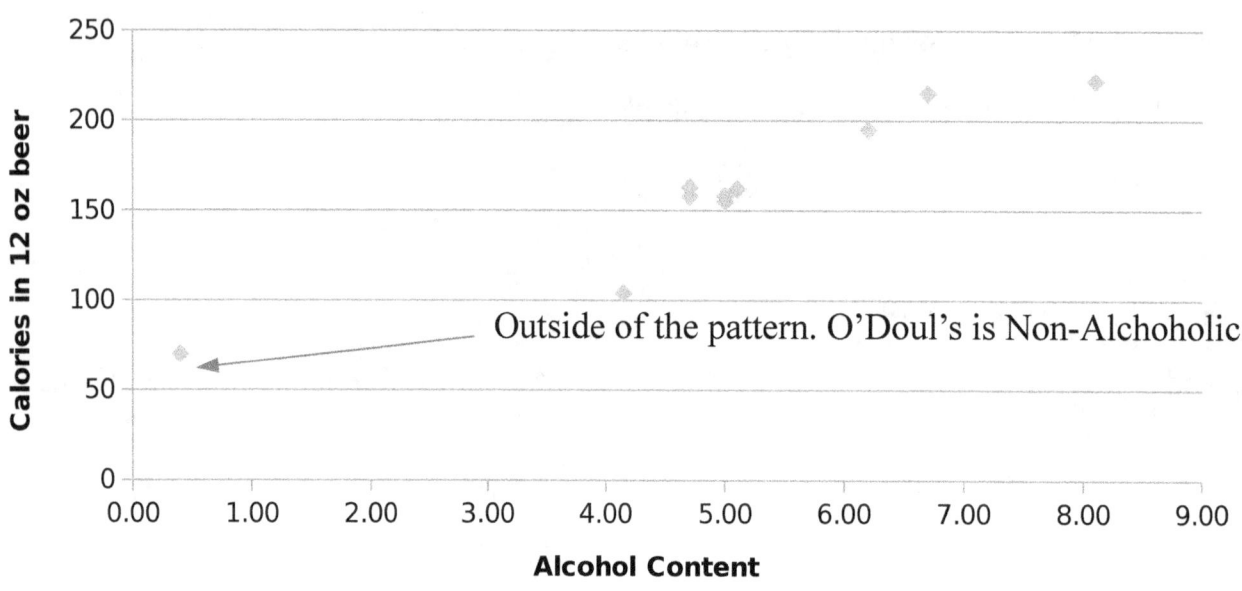

**Calories vs Alcohol Content**

Calories in 12 oz beer

Outside of the pattern. O'Doul's is Non-Alchoholic

Alcohol Content

Regression line and Residuals

- Start by drawing a line that goes through the average point $(\bar{x}, \bar{y})$

- For each point not on the line, there is a "residual," which is the vertical distance from that point to the line.

- Now adjust the slope until the sum of squared residuals is as small as possible.

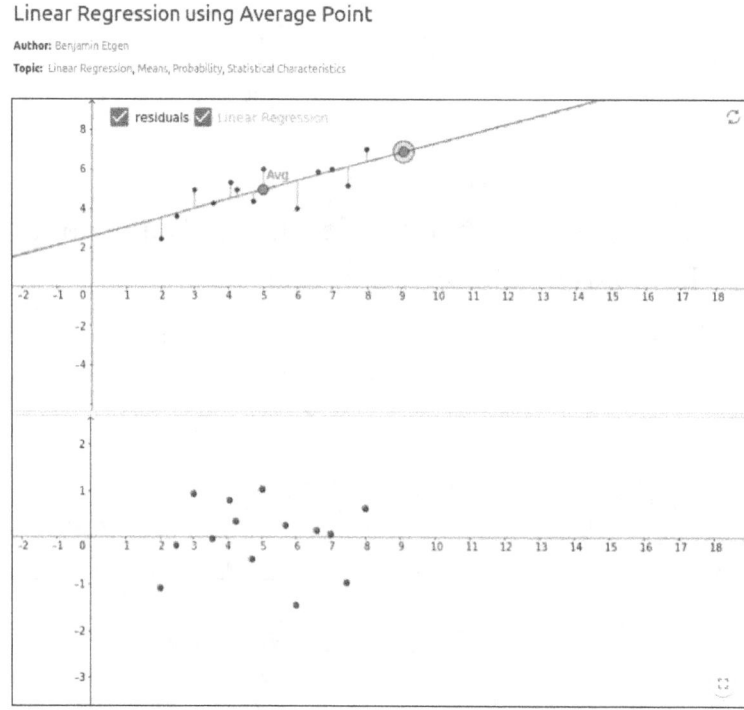

Linear Regression using Average Point

**Author:** Benjamin Etgen

**Topic:** Linear Regression, Means, Probability, Statistical Characteristics

Select "Linear" for Regression Model:

$y = ax + b$, where $a =$ slope and $b =$ y-intercept.

## The two variables and their names

- **Independent variable**, or **explanatory variable**, or **predictor variable**, is the $x$-value
- **Dependent variable**, or **response variable**, is the $y$-variable
- Sometimes there's no distinction
- If you control one variable directly, intending to thus indirectly control the other variable, the one you directly control is the independent variable.
- If time is one of the variables, we always make it the independent variable.

## Assumptions

a)  The set $(x, y)$ of ordered pairs is a random sample from the population of all such possible $(x, y)$ pairs.

b)  It is difficult to determine if the points have a normal distribution.

i) Look to see if the scatter plot has a linear pattern.

ii) Remove any data points that appear to be outliers and redo the analysis to see if the removal of the points improves the regression. Close the data analysis and preform a Multiple Variable Analysis to look for outliers. Remove outliers and redo the Two Variable Regression Analysis. (We are adding this step. It is not in the text.)

iii) Select Residual Plot to see if there is randomness in the residuals. If there is a pattern to the residuals, then there is an issue in the data.

Later, we will also use the linear correlation coefficient, $r$, to tell us if there is a linear relationship.

Finding an outlier:

Select both columns (A and B) and preform Multiple Variable Analysis.

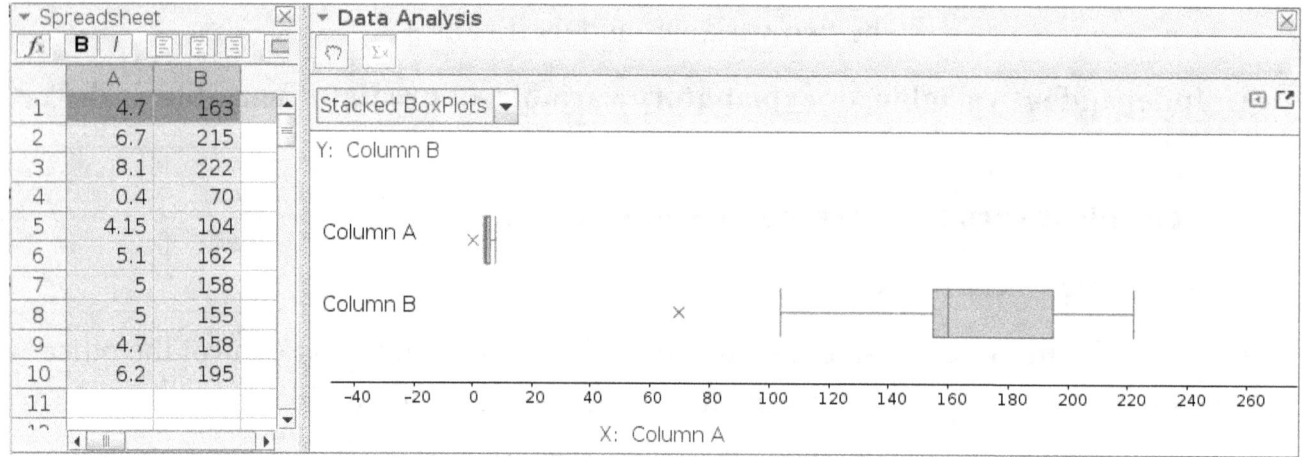

The non-alcoholic beer is an outlier. Delete it.

Select both columns (A and B) and preform Two Variable Regression Analysis.

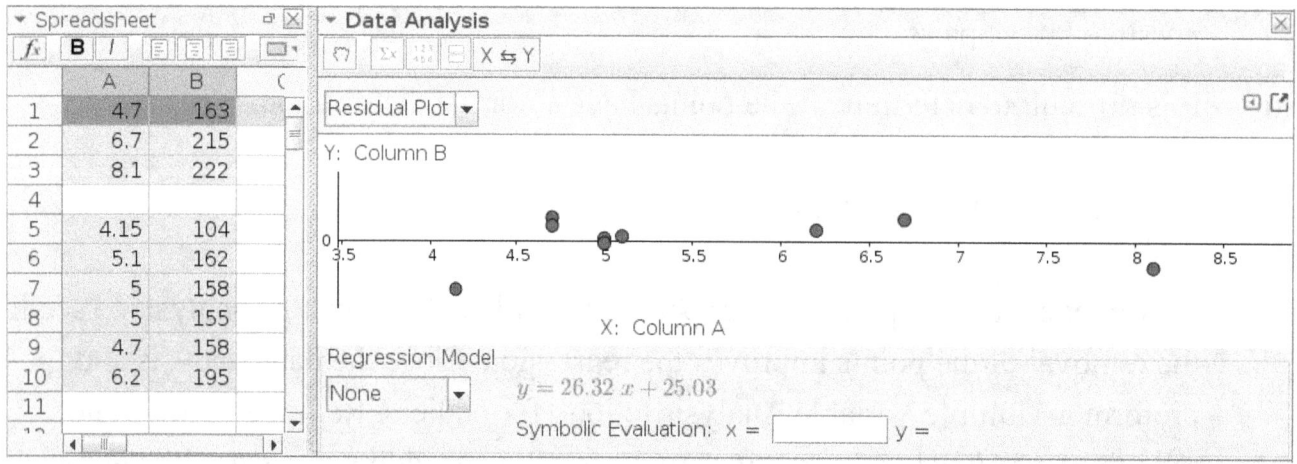

Interpreting the display

- The slope is multiplied by $x$, the intercept is added.

- $y = 26.32\,x + 25.03$ in familiar $y = mx + b$ format

Ex.: Use a random sample of different beer's alcohol content and calories to find the linear regression model for the alcohol content and the number of calories in a 12-ounce beer.

1) State the random variable


2) State and check the assumptions

a)

b)

    i)

    ii)


3) Compute the linear regression line.




Interpolation

Ex: Use the regression equation to find the number of calories in a beer when the alcohol content is 6.50%.


- $y = 26.32(6.50) + 25.03 = 196$ calories

- You can also enter 6.5 for $x$ as the Symbolic Evaluation.

- If you are drinking a beer that is 6.50% alcohol content, then it is probably close to 196 calories.

- Notice, the mean number of calories is 170 calories. This value of 196 seems like a better estimate than the mean when looking at the original data. The ⬭ equation is a better ⬭ than the ⬭.

## Extrapolation

Ex:   Use the regression equation to find the number of calories when the alcohol content is 2.00%.

- $y = 26.32(2) + 25.03 = 78$ calories

- You can also enter 2 for $x$ as the Symbolic Evaluation.

- This says that a beer that is 2.00% alcohol content probably has 78 calories. This doesn't seem like a very good estimate. This estimate is what is called extrapolation. It is not a good idea to predict values that are far outside the range of the original data. This is because you can never be sure that the regression equation is valid for data outside the original data.

## Interpret slope and intercept

- Slope is the rate of change of the dependent variable compared to the independent variable. In this case, it is the number of calories you expect to add every time you add 1% to the alcohol content.

- The intercept is what you'd expect to get for the y-variable if $x$ is zero. Since 0 is far outside the range of our data (extrapolation), that is not meaningful in this example.

## Dangers of extrapolation

When you use regression to come up with an equation to predict the growth of a city, like Flagstaff, you can run into the dangers of 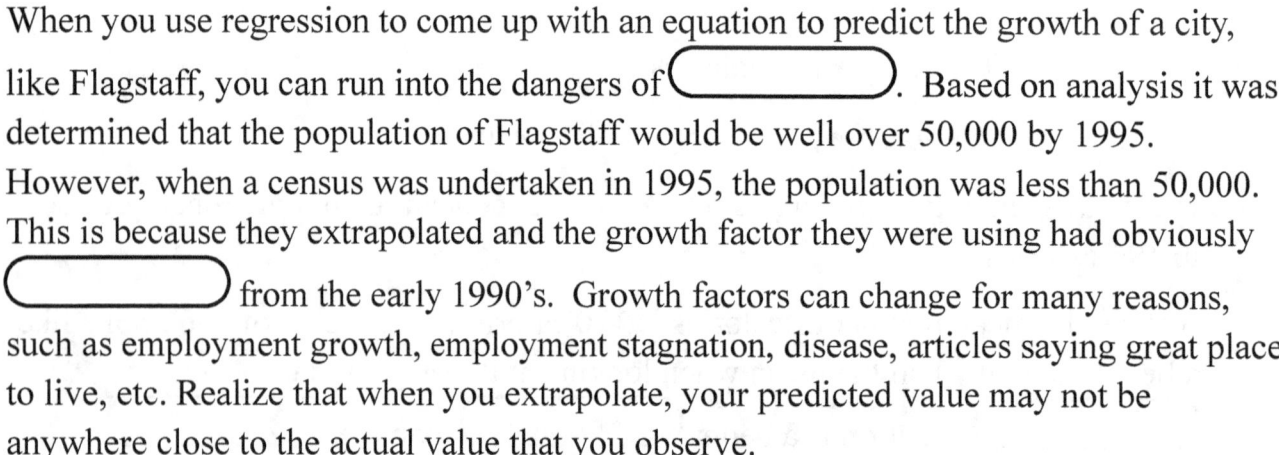. Based on analysis it was determined that the population of Flagstaff would be well over 50,000 by 1995. However, when a census was undertaken in 1995, the population was less than 50,000. This is because they extrapolated and the growth factor they were using had obviously _____ from the early 1990's. Growth factors can change for many reasons, such as employment growth, employment stagnation, disease, articles saying great place to live, etc. Realize that when you extrapolate, your predicted value may not be anywhere close to the actual value that you observe.

Ex: When an anthropologist finds skeletal remains, they need to figure out the height of the person. The height of a person (in cm) and the length of their metacarpal bone 1 (in mm) were collected. Create a scatter plot and find a regression equation between the height of a person and the length of their metacarpal. Then use the regression equation to find the height of a person for a metacarpal length of 44 mm and for a metacarpal length of 55 mm. Which height that you calculated do you think is closer to the true height of that person? Why?

| Length of Metacarpal (mm) | Height of Person (cm) |
|---|---|
| 45 | 171 |
| 51 | 178 |
| 39 | 157 |
| 41 | 163 |
| 48 | 172 |
| 49 | 183 |
| 46 | 173 |
| 43 | 175 |
| 47 | 173 |

## Section 10.2: Correlation

- A **correlation** exists between two variables when the values of one variable are somehow ◯⎯⎯⎯◯ with the values of the other variable.

- When you see a pattern in the data you say there is a ◯⎯⎯⎯◯ in the data. Though this book is only dealing with linear patterns, patterns can be exponential, logarithmic, or periodic. To see this pattern, you can draw a scatter plot of the data.

- The words "weak", "moderate", and "strong" are used to describe the ◯⎯⎯⎯◯ of the relationship between the two variables.

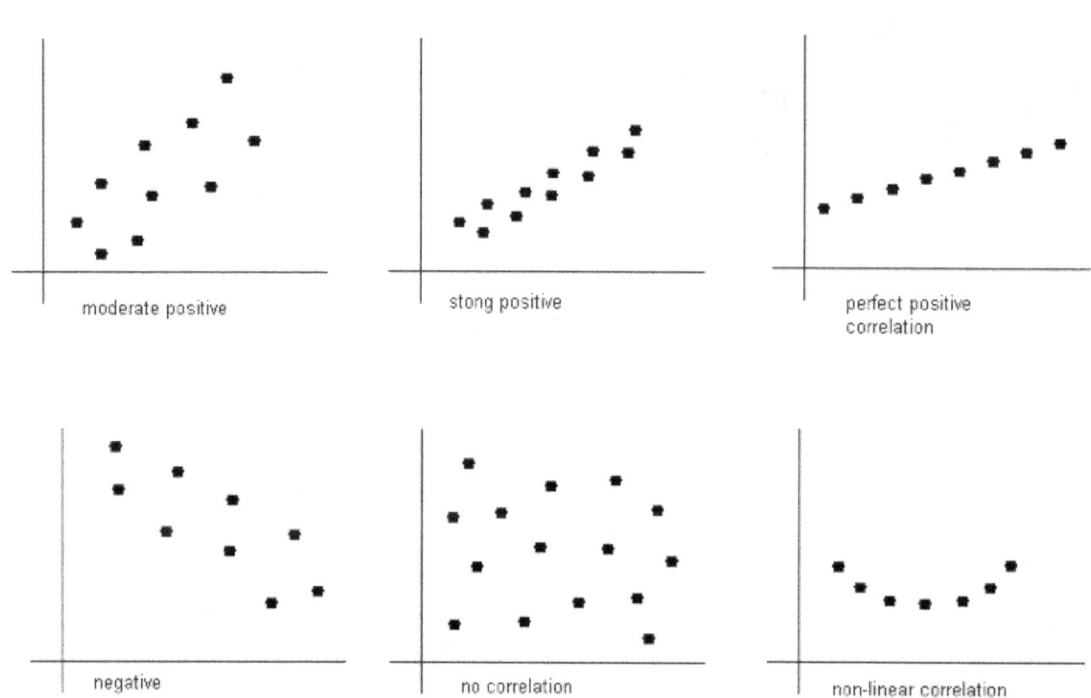

moderate positive          stong positive          perfect positive
                                                   correlation

negative          no correlation          non-linear correlation

116

## Linear correlation coefficient

The linear correlation coefficient is a number that describes the $\bigcirc$ of the linear relationship between the two variables.  It is also called the $\bigcirc$ correlation coefficient after Karl Pearson who developed it.  The symbol for the sample linear correlation coefficient is *r*.  The symbol for the population correlation coefficient is $\rho$ (Greek letter rho).

## Interpretation of the correlation coefficient

*r* is always between -1 and 1.  If *r* = -1, then there is a perfect negative (downward) linear correlation and *r* = 1 means there is a perfect positive (upward) correlation.  The closer *r* is to 1 or -1, the $\bigcirc$ the correlation.  The closer *r* is to 0, the $\bigcirc$ the correlation.  CAREFUL:  *r* = 0 does not mean there is no correlation.  It just means there is **no linear correlation.**  There might be a very strong curved pattern.

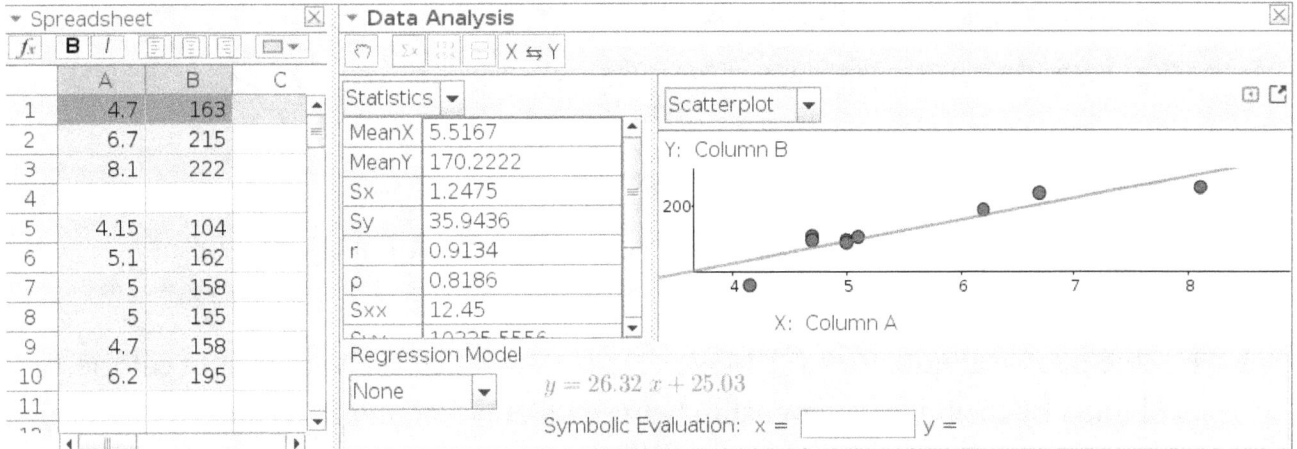

## Causation

- Correlation does not prove $\bigcirc$.

- However, lack of correlation pretty much rules out $\bigcirc$.

- It is very difficult to establish $\bigcirc$ by observational data like regression analysis.

- Properly designed $\bigcirc$ establish causation.

- Sometimes, $\bigcirc$ is all we can do.

Ice cream and drownings

- There is a very strong positive correlation between ice cream consumption and drowning deaths.

- 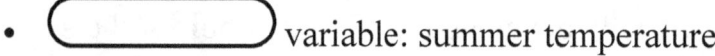 variable: summer temperature

## Explained Variation

(total variation) = (explained variation) + (unexplained variation)

The proportion of the variation that is explained by the model is:

$$r^2 = \frac{\text{explained variation}}{\text{total variation}}$$

This is known as the **coefficient of determination**.

## r and $r^2$

- In a linear correlation, $r^2$ is just the square of $r$

- For nonlinear correlations, there is no $r$. But there is still a coefficient of determination $r^2$

- For the beer example, $r^2 \approx 0.834375$. Thus, 83.44% of the variation in calories is explained by the linear relationship between alcohol content and calories. The other 16.56% of the variation is due to other factors.

Ex:  Use a random sample of different beer's alcohol content and calories to find the correlation coefficient and coefficient of determination and then interpret both.

Ex:  When an anthropologist finds skeletal remains, they need to figure out the height of the person. Use the table to find the correlation coefficient and coefficient of determination and then interpret both.

## Section 10.3: Inference for Regression & Correlation

- How do you really say you have a correlation? Can you test to see if there really is a correlation? Of course, the answer is yes.

- We can use a Hypothesis Test for Correlation.

Hypothesis Test for Correlation

1) State variables. $x$ is independent, $y$ is dependent.

2) Hypotheses & $\alpha$

$H_0$: $\rho = 0$    (There is no correlation)

$H_a$: $\rho < 0$   or   $\rho \neq 0$   or   $\rho > 0$

3) Check assumptions (same as regression)

4) Test statistic and p-value

$$t = \frac{r}{\sqrt{\dfrac{1-r^2}{n-2}}}, \quad df = n-2$$

p-value is tail or tails as usual

5) Conclusion. As always, if p-value $< \alpha$, reject the null hypothesis in favor of the alternate. There is sufficient evidence to support the alternate hypothesis. Otherwise, there is not enough evidence to support the alternate hypothesis at the stated level $\alpha$.

6) Interpretation: what does this conclusion imply in the context of the problem?

$$\bar{x} = \frac{\sum x}{n} \text{ and } \bar{y} = \frac{\sum y}{n}, \text{ calculates the average point, } (\bar{x}, \bar{y}).$$

$$b = \frac{SS_{xy}}{SS_x}, \text{ where } SS_{xy} = \sum (x - \bar{x})(y - \bar{y}) \text{ and } SS_x = \sum (x - \bar{x})^2$$

$$a = \bar{y} - b\bar{x}$$

$$\hat{y} = a + bx, \text{ where } b = \text{slope and } a = \text{y-intercept}.$$

When you compare $\hat{y} = a + bx$ with $y = mx + b$, $a$ is what we had called $b$ and $b$ is what we had called $m$.

For a hypothesis test we also need variables for the population.

$$y = \beta_0 + \beta_1 x, \text{ where } \beta_1 = \text{slope and } \beta_0 = \text{y-intercept}. \ \hat{y} \text{ is used to predict } y.$$

Ex:     Is there a positive correlation between beer's alcohol content and calories? To determine if there is a positive linear correlation, a random sample was taken of beer's alcohol content and calories for several different beers. Test at the 5% level.

We preform a Two Variable Regression Analysis as before. This tells us that $r = 0.9134$. Since we removed the outlier, there are $n = 9$ data points.

| Spreadsheet | | | | | | | Data Analysis | | | |
|---|---|---|---|---|---|---|---|---|---|---|
| | A | B | C | D | E | F | Statistics | | Data | |
| 1 | 4.7 | 163 | | | | | MeanX | 5.5167 | | X: Column A | Y: Column B |
| 2 | 6.7 | 215 | | | | | MeanY | 170.2222 | ☑1 4.7 | 163.0 |
| 3 | 8.1 | 222 | | | | | Sx | 1.2475 | ☑2 6.7 | 215.0 |
| 4 | 0.4 | 70 | | | | | Sy | 35.9436 | ☑3 8.1 | 222.0 |
| 5 | 4.15 | 104 | | | | | r | 0.9134 | ☐4 0.4 | 70.0 |
| 6 | 5.1 | 162 | | | | | ρ | 0.8186 | ☑5 4.15 | 104.0 |
| 7 | 5 | 158 | | | | | Sxx | 12.45 | ☑6 5.1 | 162.0 |
| 8 | 5 | 155 | | | | | Syy | 10335.5556 | ☑7 5.0 | 158.0 |
| 9 | 4.7 | 158 | | | | | Sxy | 327.6667 | ☑8 5.0 | 155.0 |
| 10 | 6.2 | 195 | | | | | | | ☑9 4.7 | 158.0 |
| 11 | | | | | | | | | ☑10 6.2 | 195.0 |
| 12 | | | | | | | R² | 0.8344 | | |
| 13 | | | | | | | SSE | 1711.8251 | | |
| 14 | | | | | | | | | | |
| 15 | | | | | | | | | | |
| 16 | | | | | | | | | | |
| 17 | | | | | | | | | | |

We use the formula for $t$:

$$t = \frac{r}{\sqrt{\frac{1-r^2}{n-2}}} = \frac{0.9134}{\sqrt{\frac{1-0.9134^2}{9-2}}} \approx 5.936742284$$

We are testing for a positive linear correlation, so $H_a: \rho > 0$. This means we are preforming a right tail test. Since $df = n - 2$, $df = 7$.

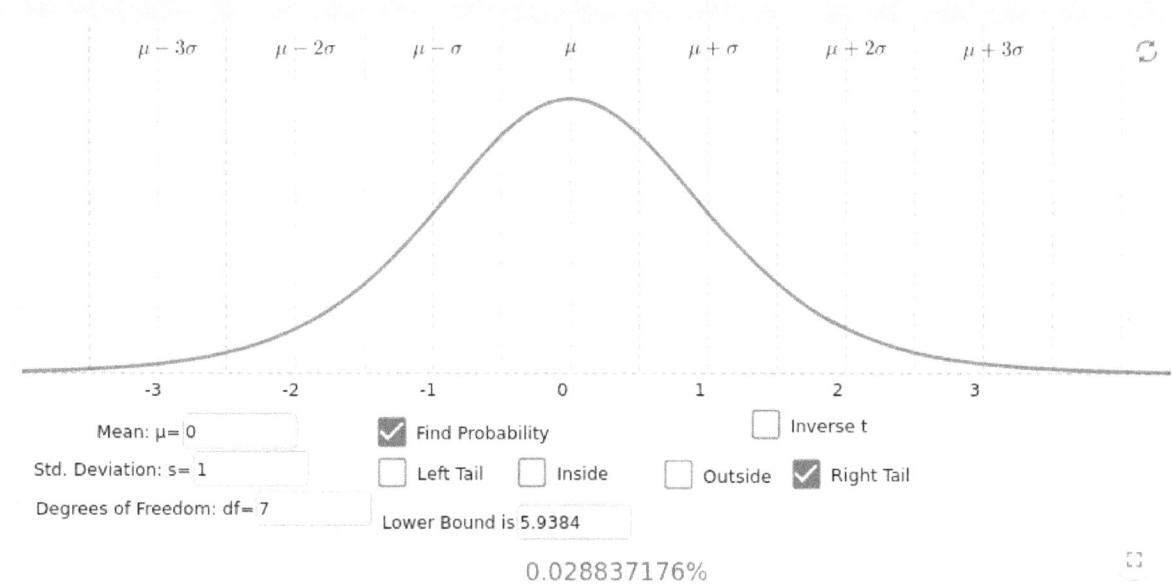

**Gosset Student T Distribution Calculator**
Showing Always

Mean: μ= 0    ☑ Find Probability    ☐ Inverse t

Std. Deviation: s= 1    ☐ Left Tail   ☐ Inside   ☐ Outside   ☑ Right Tail

Degrees of Freedom: df= 7    Lower Bound is 5.9384

0.028837176%

The p-value is 0.0002884. Since 0.0002884 < 0.05, we reject the null hypothesis in favor of the alternative. There is sufficient evidence to support the claim that there is a positive correlation between beer's alcohol content and calories.

1) State variables

2) Hypotheses & $\alpha$

3) Check assumptions
a)
b)
      i)
      ii)

4) Test statistic and p-value

5) Conclusion

6) Interpretation

It would be nice to have a range instead of a single value. The range is called a prediction interval. To find this, you need to figure out how much error is in the estimate from the regression equation. This is known as the **standard error of the estimate**.

$$s_e = \sqrt{\frac{\sum (y - \hat{y})^2}{n-2}}$$ where $\sum (y - \hat{y})^2$ is shown as the Sum of the Squared Errors, SSE.

Thus, $s_e = \sqrt{\frac{SSE}{n-2}}$.

Standard error from the beer example:

| Statistics ▼ | |
|---|---|
| MeanX | 5.5166666667 |
| MeanY | 170.2222222222 |
| Sx | 1.247497495 |
| Sy | 35.9436287045 |
| r | 0.9134413647 |
| ρ | 0.8185726876 |
| Sxx | 12.45 |
| Syy | 10335.5555555556 |
| Sxy | 327.6666666667 |
| | |
| R² | 0.8343751268 |
| SSE | 1711.8250780901 |

- $s_e = \sqrt{\frac{1711.8250780901}{9-2}} \approx 15.637980679$

Ex: When an anthropologist finds skeletal remains, they need to figure out the height of the person. The height of a person (in cm) and the length of their metacarpal bone one (in mm) were collected. Test at the 1% level for a positive correlation between length of metacarpal bone one and height of a person. Find the standard error of the estimate. (Use the table on page 115.)

Lets compare $r^2$ with $s_e$ by asking what would happen if we watered-down each of the beers. If we added an equal amount of water for each quantity of beer, both the calories and the alcohol content would fall by a half as well. (Recall that the outlier has been removed.)

Watered-down Beers:

| Alcohol Content | Calories in 12 oz |
|---|---|
| 2.35 | 81.5 |
| 3.35 | 107.5 |
| 4.05 | 111 |
| 2.075 | 52 |
| 2.55 | 81 |
| 2.5 | 79 |
| 2.5 | 77.5 |
| 2.35 | 79 |
| 3.1 | 97.5 |

| Statistics ▼ | |
|---|---|
| MeanX | 2.7583 |
| MeanY | 85.1111 |
| Sx | 0.6237 |
| Sy | 17.9718 |
| r | 0.9134 |
| ρ | 0.8186 |
| Sxx | 3.1125 |
| Syy | 2583.8889 |
| Sxy | 81.9167 |
| | |
| R² | 0.8344 |
| SSE | 427.9563 |

Notice that $r^2$ is the same as before we watered down the beers. For both the original beer and the watered-down beer, 83.44% of the variation in calories is explained by the linear relationship between alcohol content and calories. This makes sense because both the alcohol content and calories have fallen by a half.

However, the average distance between the linear regression line and the data, as measured by $s_e$, has changed.

Beers: $s_e = \sqrt{\dfrac{1711.8250780901}{9-2}} \approx 15.637980679$

Watered-down Beers: $s_e = \sqrt{\dfrac{427.9563}{9-2}} \approx 7.818990618$

The new data is half as far from the linear regression line. For this reason, we say that the standard error, $s_e$, is a better measure of how well the the linear regression line models the data.

The coefficient of determination, $r^2$, tells us what proportion of the variation that is explained by the model.

# Chapter 11

## Chi-Square

### Section 11.1: Chi-Square Test for Independence

- Two categorical variables

- Are they independent, or are they related?

- If they are independent, there can not be any $\bigcirc\!\!\!\!\!\!\!\!\!\!\!\!\!\!\!\bigcirc$ relationship.

### Chi-Square test for independence

1) Hypotheses & $\alpha$

   $H_0$: the two variables are independent

   $H_a$: the two variables are dependent

2) Assumptions: random sample, expected frequencies all $\geq 5$

3) Test statistic and p-value

   The test statistic, $\chi^2$, can be used to calculate a p-value

4) Conclusion. As always, if p-value $< \alpha$, reject the null hypothesis in favor of the alternate. There is sufficient evidence to support the alternate hypothesis that the categorical variables are dependent. Otherwise, there is not enough evidence to support the alternate hypothesis at the stated level $\alpha$.

5) Interpretation: what does this conclusion imply in the context of the problem?

### The test statistic $\chi^2$

- This is a single number that captures the central question: how $\bigcirc\!\!\!\!\!\!\!\!\!\!\!\!\!\!\!\bigcirc$ is the observed data away from what we'd expect if the null hypothesis is true?

- It is just the sum of the $\bigcirc\!\!\!\!\!\!\!\!\!\!\!\!\!\!\!\bigcirc$ of the differences but normalized to become a unit-less number.

Observations of two categorical variables for survivors of the Titanic:

| Observed Counts | | | |
|---|---|---|---|
| Class | Gender | | Total |
| | Female | Male | |
| 1st | 134 | 59 | 193 |
| 2nd | 94 | 25 | 119 |
| 3rd | 80 | 58 | 138 |
| Total | 308 | 142 | 450 |

What are the expected values?

| Class | Gender | | Total |
|---|---|---|---|
| | Female | Male | |
| 1st | | | 193 |
| 2nd | | | 119 |
| 3rd | | | 138 |
| Total | 308 | 142 | 450 |

Focus on each cell one at a time, and ask "what should we expect if gender is independent of class?"

Focus on "3rd Class" and "Female" cell

- Among the 450 survivors, 308 were women. This works out to 68.4%.

- If there were no relationship between gender and class, that same percentage of the 3rd class survivors would be expected to be women. 68.4% of 138 is 94.45.

- Among the 450 survivors, 138 were from 3rd class. This works out to 30.7%.

- If there were no relationship between gender and class, that same percentage of the female survivors would be expected to be from 3rd class. 30.7% of 308 is again 94.45.

For each cell, the expected count is:

$$\frac{(\text{row total})\cdot(\text{column total})}{\text{grand total}}$$

| Expected Counts | | | |
|---|---|---|---|
| Class | Gender | | Total |
| | Female | Male | |
| 1st | 132.0978 | 60.9022 | 193 |
| 2nd | 81.4489 | 37.5511 | 119 |
| 3rd | 94.4533 | 43.5467 | 138 |
| Total | 308 | 142 | 450 |

How "far apart" are these two matrices?

- In each cell, ⬭ to calculate a distance.

- Since we're going to want to add them up to get a grand total distance, we need to ⬭ each difference first, to make all the numbers positive, just as we have done before in things like the standard deviation.

- We ⬭ each square by the expected number first. This makes the resulting sum unit-less. Finally, we add them up

$$\chi^2 = \sum \frac{(\text{observed}_{ij} - \text{expected}_{ij})^2}{\text{expected}_{ij}}$$

For this data, $\chi^2 \approx 13.22$

*Getting to the p-value*

- The area of a right tail of The $\chi^2$ distribution function is the p-value.

- Actually, like the t-distribution, there is a whole family of $\chi^2$ functions, one for each degree of freedom.

- The degree of freedom is defined as: (row count-1)×(column count -1)

*Shape of $\chi^2$ distribution*

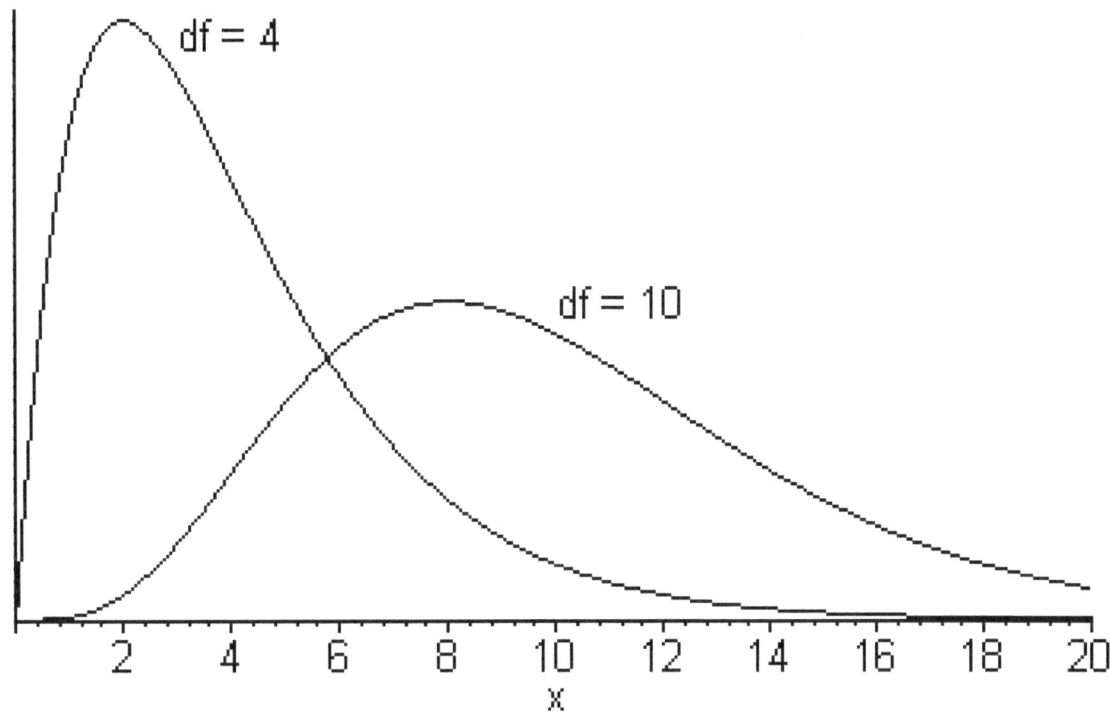

Using the Chi Squared Test: p-value $\approx 0.0013468321$

- That is less than 5%, which is a reasonable $\alpha$ for this problem. So we reject the null hypothesis.

- There is enough evidence to show that the class and gender of survivors of the Titanic are dependent.

| Observed Counts | | | |
|---|---|---|---|
| Class | Gender | | Total |
| | Female | Male | |
| 1st | 134 | 59 | 193 |
| 2nd | 94 | 25 | 119 |
| 3rd | 80 | 58 | 138 |
| Total | 308 | 142 | 450 |

| Expected Counts | | | |
|---|---|---|---|
| Class | Gender | | Total |
| | Female | Male | |
| 1st | 132.0978 | 60.9022 | 193 |
| 2nd | 81.4489 | 37.5511 | 119 |
| 3rd | 94.4533 | 43.5467 | 138 |
| Total | 308 | 142 | 450 |

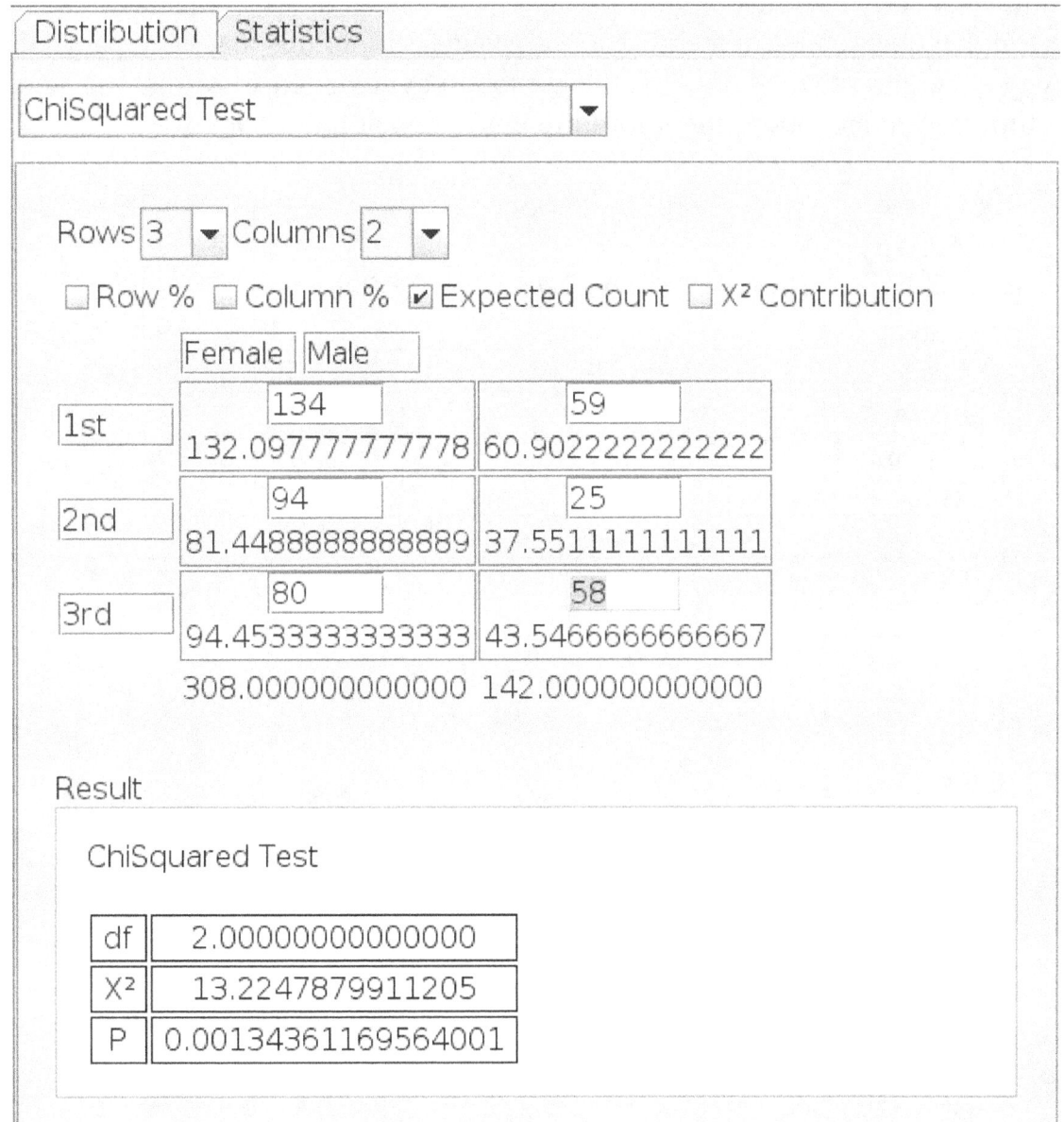

Distribution | Statistics

ChiSquared Test

Rows 3 ▼ Columns 2 ▼

☐ Row % ☐ Column % ☑ Expected Count ☐ X² Contribution

| | Female | Male |
|---|---|---|
| 1st | 134<br>132.097777777778 | 59<br>60.9022222222222 |
| 2nd | 94<br>81.4488888888889 | 25<br>37.5511111111111 |
| 3rd | 80<br>94.4533333333333 | 58<br>43.5466666666667 |
| | 308.000000000000 | 142.000000000000 |

Result

ChiSquared Test

| df | 2.00000000000000 |
|---|---|
| X² | 13.2247879911205 |
| P | 0.00134361169564001 |

Ex:    Researchers watched groups of dolphins off the coast of Ireland in 1998 to determine what activities the dolphins partake in at certain times of the day.  The numbers in the table represent the number of groups of dolphins that were partaking in an activity at certain times of days.  Is there enough evidence to show that the activity and the time period are independent for dolphins?  Test at the 1% level.

| Activity | Period | | | | Row Total |
|---|---|---|---|---|---|
| | Morning | Noon | Afternoon | Evening | |
| Travel | 6 | 6 | 14 | 13 | 39 |
| Feed | 28 | 4 | 0 | 56 | 88 |
| Social | 38 | 5 | 9 | 10 | 62 |
| Column Total | 72 | 15 | 23 | 79 | 189 |

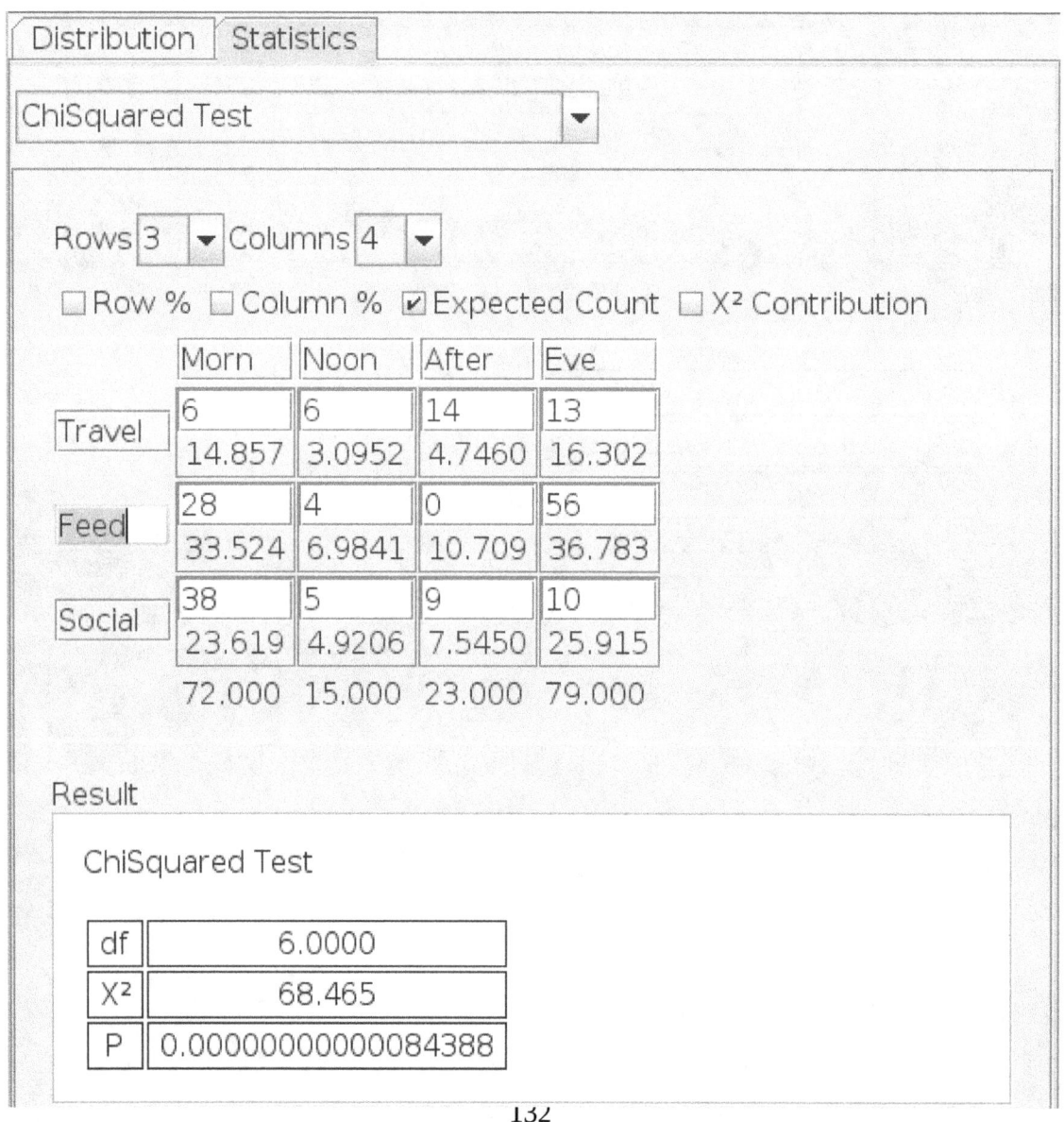

Distribution | Statistics

ChiSquared Test

Rows 3 ▼ Columns 4 ▼

☐ Row %  ☐ Column %  ☑ Expected Count  ☐ X² Contribution

| | Morn | Noon | After | Eve |
|---|---|---|---|---|
| Travel | 6 | 6 | 14 | 13 |
| | 14.857 | 3.0952 | 4.7460 | 16.302 |
| Feed | 28 | 4 | 0 | 56 |
| | 33.524 | 6.9841 | 10.709 | 36.783 |
| Social | 38 | 5 | 9 | 10 |
| | 23.619 | 4.9206 | 7.5450 | 25.915 |
| | 72.000 | 15.000 | 23.000 | 79.000 |

Result

ChiSquared Test

| df | 6.0000 |
|---|---|
| X² | 68.465 |
| P | 0.0000000000084388 |

1) Hypotheses & $\alpha$

   $H_0$: the activity and the time period are independent for dolphins

   $H_a$: the activity and the time period are dependent for dolphins

   $\alpha = 0.01$

2) Assumptions

   i. A sample of the activity and the time period for dolphins is not known to be random.

   ii. Expected frequencies for each cell are greater than or equal to 5. Not all frequencies are more than 5. There are two that are below 5, so this assumption is not true and the results of the hypothesis test may not be valid.

3) Test statistic and p-value

   $\chi^2 \approx 68.465$

   p-value $= 8.4388 \times 10^{-13}$

4) Conclusion. Since p-value $< \alpha$, reject the null hypothesis in favor of the alternate.

5) There is enough evidence to show that the activity and the time period are dependent for dolphins.

## Section 11.2: Chi-Square Goodness of Fit

- Sometimes, we want to test a single categorical value to see if the observed values match theoretical values.

- For example, a casino wants to make sure roulette wheels and dice are "fair". As long as any departures from "fairness" are much smaller than the "house edge," the gambling equipment is safe to use.

### Babies Birthdays

| Day | Sunday | Monday | Tuesday | Wednesday | Thursday | Friday | Saturday |
|---|---|---|---|---|---|---|---|
| 50 random | 7 | 11 | 8 | 9 | 4 | 5 | 6 |
| 300 random | 33 | 39 | 54 | 43 | 45 | 43 | 43 |
| 1300 random | 149 | 206 | 207 | 183 | 178 | 203 | 174 |

- Do the weekend numbers seem smaller than the others?

Let's assume that the chances of being born on any day is the same: 1/7. For 50 random babies selected, we would expect that $50 \div 7 \approx 7.1429$ babies were born each day.

| Day | Sunday | Monday | Tuesday | Wednesday | Thursday | Friday | Saturday |
|---|---|---|---|---|---|---|---|
| Observed | 7 | 11 | 8 | 9 | 4 | 5 | 6 |
| Expected | 7.1429 | 7.1429 | 7.1429 | 7.1429 | 7.1429 | 7.1429 | 7.1429 |

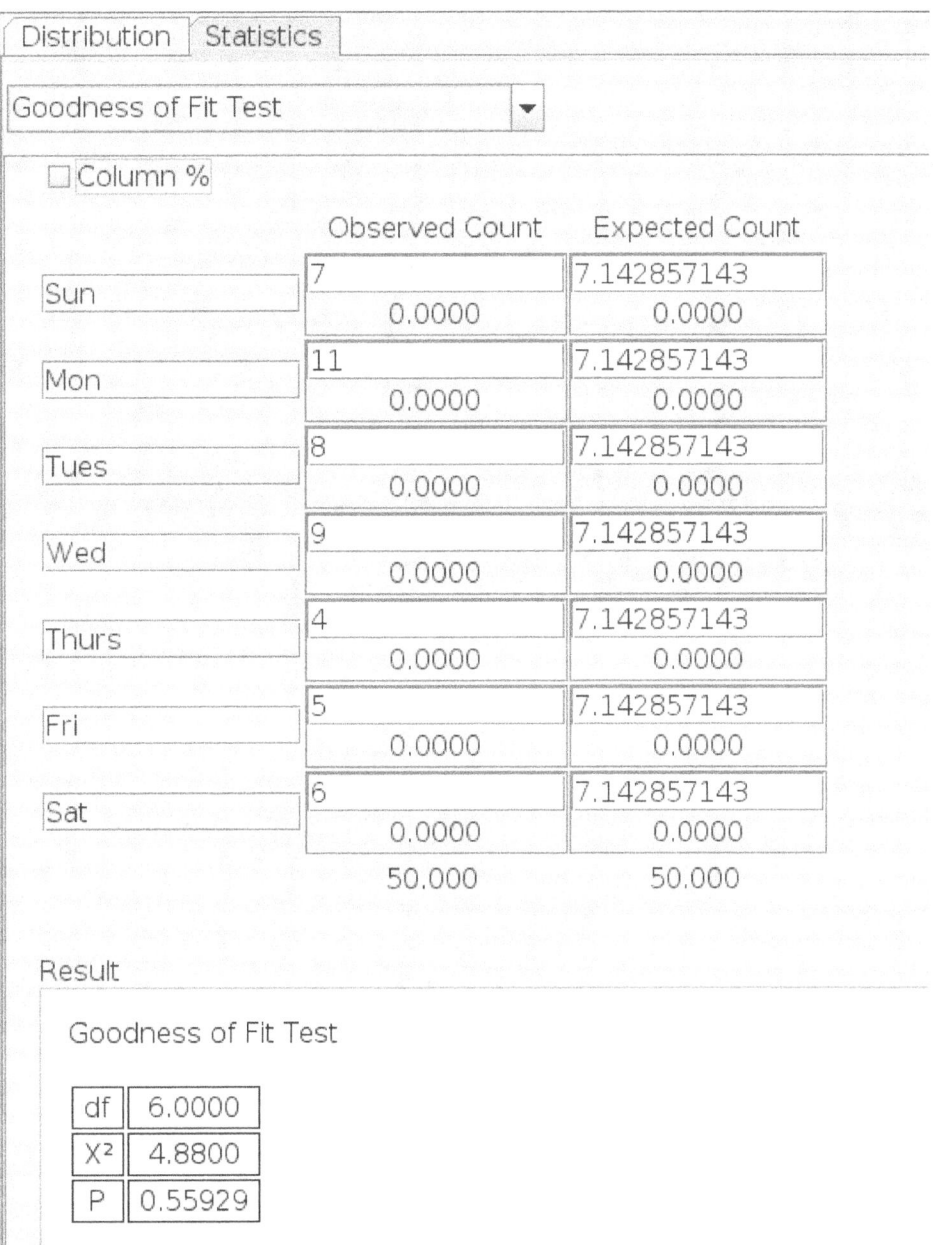

Goodness of Fit Test

| | Observed Count | Expected Count |
|---|---|---|
| Sun | 7 | 7.142857143 |
| | 0.0000 | 0.0000 |
| Mon | 11 | 7.142857143 |
| | 0.0000 | 0.0000 |
| Tues | 8 | 7.142857143 |
| | 0.0000 | 0.0000 |
| Wed | 9 | 7.142857143 |
| | 0.0000 | 0.0000 |
| Thurs | 4 | 7.142857143 |
| | 0.0000 | 0.0000 |
| Fri | 5 | 7.142857143 |
| | 0.0000 | 0.0000 |
| Sat | 6 | 7.142857143 |
| | 0.0000 | 0.0000 |
| | 50.000 | 50.000 |

Result

Goodness of Fit Test

| df | 6.0000 |
|---|---|
| $X^2$ | 4.8800 |
| P | 0.55929 |

50 babies data interpreted:

$\chi^2$ = 4.880, p-value = 0.55929

- This means there is a 56% chance of getting data like this if babies are equally likely to be born on any weekday.

- There is insufficient evidence that baby's birth days are not equally likely on any day of the week.

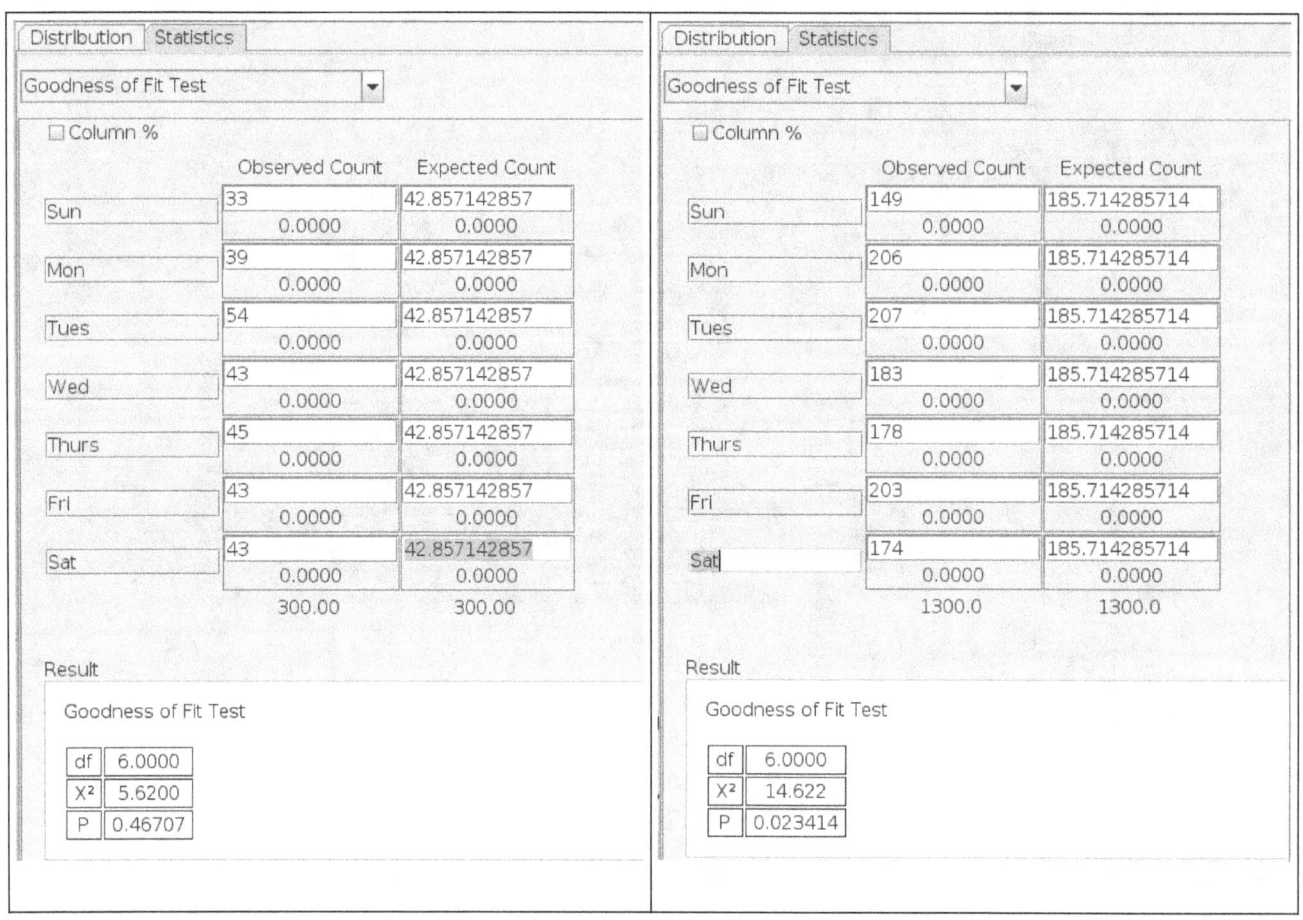

300 baby data: $\chi^2 = 5.62$, p-value = 0.467

Still not much evidence. This could happen by chance, easily.

1300 baby data: $\chi^2 = 14.6$, p-value = 0.0234

"With 1300 randomly-selected babies, we have evidence that babies birth days were not uniformly distributed among the weekdays."

Ex  According to the M&M candy company, the expected proportion can be found in the table. In addition, the table contains the number of M&M's of each color that were found in a case of candy. At the 5% level, do the observed frequencies support the claim of M&M?

**M&M Observed and Proportions**

|  | Blue | Brown | Green | Orange | Red | Yellow | Total |
|---|---|---|---|---|---|---|---|
| Observed Frequencies | 481 | 371 | 483 | 544 | 372 | 369 | 2620 |
| Expected Proportion | 0.24 | 0.13 | 0.16 | 0.20 | 0.13 | 0.14 | |
| Expected | | | | | | | |

1)  Hypotheses & $\alpha$

$H_0$:

$H_a$:

$\alpha = 0.05$

2)  Assumptions

i.

ii.

3)  Test statistic and p-value

4)  Conclusion.

5)

## Section 11.3: Analysis of Variance (ANOVA)

- We use analysis of variance (ANOVA) to compare three or more means.
- Since the test statistic for ANOVA is complicated, we will use technology to find the test statistic and p-value.
- To obtain a statistically significant result there need only be a difference between any two of the means.

Hypothesis test using ANOVA to compare $k$ means

1) State the random variables and the parameters in words

   $X_1$ =random variable 1

   $X_2$ =random variable 2

   $\vdots$

   $X_k$ =random variable k

   $\mu_1$ =mean of random variable 1

   $\mu_2$ =mean of random variable 2

   $\vdots$

   $\mu_k$ =mean of random variable k

2) State the null and alternative hypotheses and the level of significance

   $H_0: \mu_1 = \mu_2 = \mu_3 = \cdots = \mu_k$

   $H_a$: at least two of the means are not equal

   Also state $\alpha$

3) State and check the assumptions for the hypothesis test

   i       A random sample of size $n_i$ is taken from each population.

   ii      All the samples are independent of each other.

   iii     The ANOVA test is reliable if the sample sizes are fairly close to each other.

4) Find the test statistic and p-value

   Enter the data and preform a Multiple Variable Analysis. Select ANOVA.

5) Conclusion. As always, if p-value $< \alpha$, reject the null hypothesis in favor of the alternate. There is sufficient evidence to support the alternate hypothesis that the categorical variables are dependent. Otherwise, there is not enough evidence to support the alternate hypothesis at the stated level $\alpha$.

6) Interpretation: what does this conclusion imply in the context of the problem?

Ex    The amount of sodium (in mg) in different types of hotdog is in the table.  Is there sufficient evidence to show that the mean amount of sodium in the types of hotdog are not equal?  Test at the 5% level.

**Amount of Sodium (in mg) in Beef, Meat, and Poultry Hotdogs**

| Beef | Meat | Poultry |
|------|------|---------|
| 495 | 458 | 430 |
| 477 | 506 | 375 |
| 425 | 473 | 396 |
| 322 | 545 | 383 |
| 482 | 496 | 387 |
| 587 | 360 | 542 |
| 370 | 387 | 359 |
| 322 | 386 | 357 |
| 479 | 507 | 528 |
| 375 | 393 | 513 |
| 330 | 405 | 426 |
| 300 | 372 | 513 |
| 386 | | |
| 401 | | |

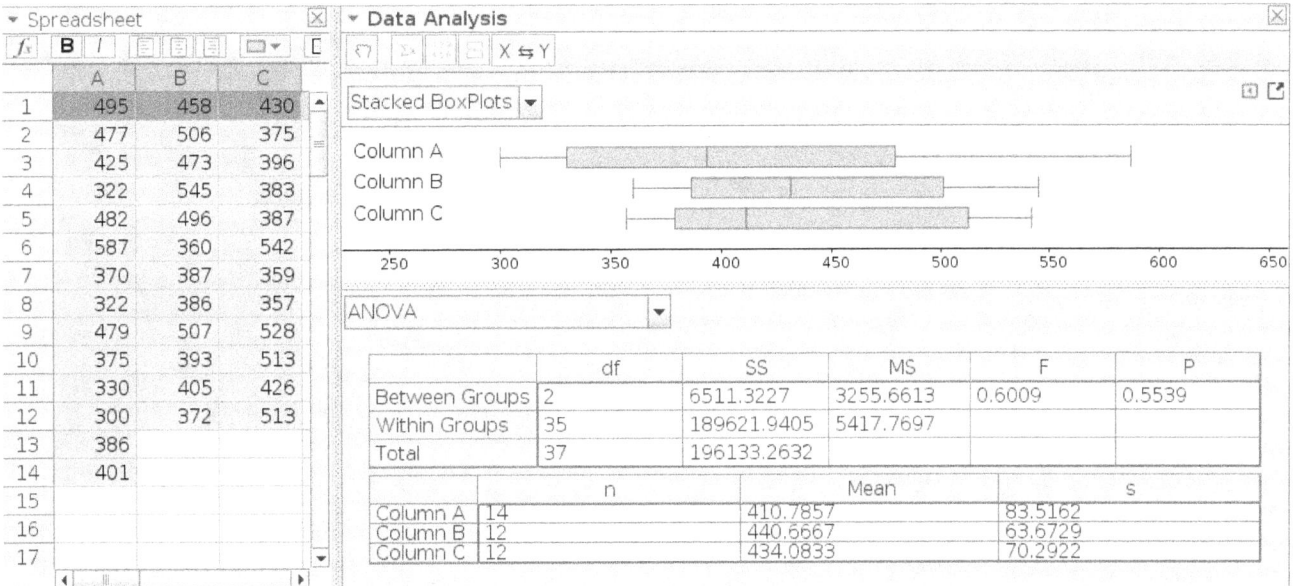

1)    State the random variables and the parameters in words

139

$x_1$ = amount of sodium in a beef hotdog
$x_2$ = amount of sodium in a meat hotdog
$x_3$ = amount of sodium in a poultry hotdog
$\mu_1$ = mean amount of sodium in beef hotdogs
$\mu_2$ = mean amount of sodium in meat hotdogs
$\mu_3$ = mean amount of sodium in poultry hotdogs

2) State the null and alternative hypotheses and the level of significance
$H_0: \mu_1 = \mu_2 = \mu_3$
$H_a$: at least two of the means are not equal
$\alpha = 0.05$

3) State and check the assumptions for the hypothesis test
i

ii

iii

4) Find the test statistic and p-value

5)

6)

*Warm-up problems*:

1    The brother of the chief executive officer (CEO) of a major industrial firm died. The man who died had no brother. How is this possible?

2    Some months have 30 days and some have 31 days. How many months have 28 days?

*Mathematical Thinking*:

3    A lilly pad grows so that each day it doubles its size. On the twentieth day of its life, it completely covers a pond. On which day did it cover half the pond?

4    A frog is at the bottom of a 20-ft well. Each day the frog crawls up 4 ft but each night the frog slips back 3 ft. After how many days will the frog reach the top of the well?

5    Two chipmunks, Chip and Dale, collect 32 acorns every day and add them to their acorn stash. The first day, after Chip fell asleep, Dale ate half the acorns from the stash. This continued for three more days until, on the morning of the fifth day, when Dale counted the stash, there were only 35. How many acorns had they started with on the first day?

Imagine that you are tossing a coin and record the sequence of 60 tosses. Use "H" for heads and "T" for tails:

| | | | | | | | | | | | | |
|---|---|---|---|---|---|---|---|---|---|---|---|---|
| | | | | | | | | | | | | |
| | | | | | | | | | | | | |
| | | | | | | | | | | | | |

Actually tossing a coin and record the sequence of 60 tosses. Use "H" for heads and "T" for tails:

| | | | | | | | | | | | | |
|---|---|---|---|---|---|---|---|---|---|---|---|---|
| | | | | | | | | | | | | |
| | | | | | | | | | | | | |
| | | | | | | | | | | | | |

A "run" is a sequence of the same face in a row. The length of a run is how many of the same face are in a row. For example, HHH is a run of three heads in a row. Compare the number of runs and the length of the runs for the imaginary coin toss with the actual coin toss.

**Name:**

Consider the sequence of heads and tails.

| H | T | H | H | T | H | T | T | H | T | H | H | T | H | T |
|---|---|---|---|---|---|---|---|---|---|---|---|---|---|---|
| T | H | H | T | H | T | T | H | T | H | H | T | H | T | T |
| H | H | T | H | T | T | H | T | H | H | T | H | T | H | H |
| T | H | H | T | H | T | H | T | H | H | T | H | T | T | H |

Although this sequence appears to be random, there are is a prominent reason why it may not be. Explain why this sequence of heads and tails is probably not random.

The Pythagoreans distinguished ten different "averages". We will consider two of those averages that are still used today. The "average quantity" is the sum of the quantities divided by their number, $\bar{x} = \dfrac{\sum x}{n}$, where $x$ are the quantities and $n$ is the number. The "average quantity" of 0.50 and 1.50 is $\bar{x} = \dfrac{0.5 + 1.5}{2} = \dfrac{2}{2} = 1$.

Ex 1:  Now suppose that we wanted to find the "average rate." Suppose that you invested \$1000. In the first year the value of the investment went down by 50% and in the second year the value increased by 50%. That is, in the first year the investment lost 50%. We show this as -50% or -0.5. The growth multiplier is $m = 1 + r$. In the first year the growth multiplier is $m_1 = 1 + (-0.5) = 0.5$. In the second year the growth multiplier is $m_2 = 1 + 0.5 = 1.5$. What is the ending value of the account? (Hint: $A = 1000 \times (1 - 0.5) \times (1 + 0.5)$)

Ex 2:  Now we want to find the average growth multiplier, $\bar{m}$, for the account from example #1. That is, we what to find the growth multiplier that would have yielded the same result had we earned that rate both years. (Hint: $A = 1000 \times \bar{m} \times \bar{m}$)

Ex 3:  Use the definition of growth multiplier, $m = 1 + r$, to find the "average rate" at which the account in example #1 changed.

Ex 4: Let $m_1=1.3$, $m_2=0.2$ and $m_3=1.5$. Find $\bar{m}$ for these three years.

Ex 5: Let $m_1=1.4$, $m_2=0.1$, $m_3=1$ and $m_4=1.5$. Find $\bar{m}$ for these four years.

Ex 6: Let $m_1$, $m_2$, $m_3$, $m_4$ and $m_5$ be the growth multipliers for five years. Find the formula for $\bar{m}$.

Ex 7: Let $m_1 \cdots m_n$ be the growth multipliers for $n$ years. Find the formula for $\bar{m}$.

Ex 8: One account increased in value one year by 70% and decreased in value the next year by 70%. Find $\bar{m}$, the "average rate" of change for the account.

1       Let $m_1 \cdots m_n$ be the growth multipliers for $n$ years. State the formula for $\bar{m}$.

2       Let $m_1 = 1.6$, $m_2 = 0.3$, $m_3 = 1$ and $m_4 = 1.1$. Find $\bar{m}$ for these four years.

3       One account increased in value one year by 27% and decreased in value the next year by 27%. Find $\bar{m}$ and the "average rate" of change for the account.

A game of "Best Out Of Five" is interrupted when player A has won twice and player B has won once. The game ends when a player wins three times.

Finish the game 100 times. Flip a coin twice and record the results. A wins if you flip heads and B wins if you flip tails. Record your results this way:

| Game: | Round 4: | Round 5: | Wins: |
|-------|----------|----------|-------|
| 1 | A | | A |
| ... | | | |
| 100 | B | B | B |

Player A won how many of the games:

The probability that player A would have won the game had they continued playing is, approximately:

Player B won how many of the games:

The probability that player B would have won the game had they continued playing is, approximately:

A student created the following table and concluded that the probability of player B winning the game, had it continued, is 1/3.

| Rounds 1  2  3: | Round 4: | Round 5: | Wins: | Probability: |
|---|---|---|---|---|
| | B | B | B | 1/3 |
| A A B | A | | A | 2/3 |
| | B | A | A | |

Why is this wrong?

Make a table showing the true probabilities:

| Rounds 1  2  3: | Round 4: | Round 5: | Wins: | Probability: |
|---|---|---|---|---|
| A A B or ABA or BAA | | | | |
| | | | | |
| | | | | |
| | | | | |

What is the probability that player B would have won had they continued to play?

Show your work for full credit.

A game of "Best Out Of Five" is interrupted when player A has won twice and player B has won once. The game ends when a player wins three times.

1     A student created the following table and concluded that the probability of player B winning the game, had it continued, is 1/3.

| Rounds 1  2  3: | Round 4: | Round 5: | Wins: | Probability: |
|---|---|---|---|---|
|  | B | B | B | 1/3 |
| A A B | A |  | A | 2/3 |
|  | B | A | A |  |

Why is this wrong?

**2**    Make a table showing the true probabilities:

| Rounds 1 2 3: | Round 4: | Round 5: | Wins: | Probability: |
|---|---|---|---|---|
| A A B<br>or<br>ABA<br>or<br>BAA | | | | |
| | | | | |
| | | | | |
| | | | | |

What is the probability that player B would have won had they continued to play?

Show your work for full credit.

A game of "Best Out Of Seven" is interrupted when player A has won three times and player B has won twice. The game ends when a player wins four times.

**1**      Make a table showing the probabilities:

| Rounds 1 2 3 4 5: | Round 6: | Round 7: | Wins: | Probability: |
|---|---|---|---|---|
| | | | | |
| AAABB | | | | |
| | | | | |
| | | | | |

What is the probability that player B would have won had they continued to play?

A game of "Best Out Of Seven" is interrupted when player B has won three times and player A has won once. The game ends when a player wins four times.

**2**     Make a table showing the probabilities:

| Rounds 1 2 3 4: | Round 5: | Round 6: | Round 7: | Wins: |
|---|---|---|---|---|
| ABBB | | | | |

What is the probability that player A would have won had they continued to play?

The TV game show "Let's Make A Deal" allowed a contestant to chose from three different doors. A prize was behind one of the doors, the other two were empty. Suppose that after making a choice of one of the doors, the host congratulates you for not choosing a door without the prize. That door is opened revealing that there is no prize there. Then the host gives you a choice. You can stay with your original choice or switch to the other door that was not opened.

Play the game 200 times. In the first 100 games, stay with your original choice. In the second 100 games, switch to the other door.

`http://www.rossmanchance.com/applets/MontyHall/Monty04.html`

Summarize your results.

When the contestant stayed with your original choice, what was the probability of winning?

When the contestant switched from your original choice, what was the probability of winning?

A student created the following table and concluded that the probability that a contestant wins when staying with their original choice is 1/2.

| Door: | Probability: |
| --- | --- |
| Original Choice | 1/2 |
| Other Door | 1/2 |
| Open Door | 0 |

Why is this wrong?

Make a table showing the true probabilities:

| Door: | Individual Probability: | Combined Probability: |
| --- | --- | --- |
| 1 (Original Choice) | | |
| 2 | | |
| 3 | | |

Show your work for full credit.

The TV game show "Let's Make A Deal" allowed a contestant to chose from three different doors. A prize was behind one of the doors, the other two were empty. Suppose that after making a choice of one of the doors, the host congratulates you for not choosing a door without the prize. That door is opened revealing that there is no prize there. Then the host gives you a choice. You can stay with your original choice or switch to the other door that was not opened.

1      A student created the following table and concluded that the probability that a contestant wins when staying with their original choice is 1/2.

| Door: | Probability: |
|---|---|
| Original Choice | 1/2 |
| Other Door | 1/2 |
| Open Door | 0 |

Why is this wrong?

**2**     Make a table showing the true probabilities:

| Door: | Individual Probability: | Combined Probability: |
|---|---|---|
| 1 (Original Choice) | | |
| 2 | | |
| 3 | | |

**3**     What is the probability of winning if you stay?

**4**     What is the probability of winning if you switch?

## B. Etgen                                                   Birthday - Project

Let's take a pause to consider a famous problem in probability theory:

Suppose you have a room full of 30 people. What is the probability that there is at least one shared birthday?

Take a guess at the answer to the above problem. Was your guess fairly low, like around 10%? That seems to be the intuitive answer (30/365, perhaps?). Let's see if we should listen to our intuition. Let's start with a simpler problem, however.

### Example 1

Suppose three people are in a room. What is the probability that there is at least one shared birthday among these three people?

There are a lot of ways there could be at least one shared birthday. Fortunately there is an easier way. We ask ourselves "What is the alternative to having at least one shared birthday?" In this case, the alternative is that there are **no** shared birthdays. In other words, the alternative to "at least one" is having **none**. In other words, since this is a complementary event,

P(at least one) = 1 – P(none)

We will start, then, by computing the probability that there is no shared birthday. Let's imagine that you are one of these three people. Your birthday can be anything without conflict, so there are 365 choices out of 365 for your birthday. What is the probability that the second person does not share your birthday? There are 365 days in the year (let's ignore leap years) and removing your birthday from contention, there are 364 choices that will guarantee that you do not share a birthday with this person, so the probability that the second person does not share your birthday is 364/365. Now we move to the third person. What is the probability that this third person does not have the same birthday as either you or the second person? There are 363 days that will not duplicate your birthday or the second person's, so the probability that the third person does not share a birthday with the first two is 363/365.

We want the second person not to share a birthday with you *and* the third person not to share a birthday with the first two people, so we use the multiplication rule:

$P$(no shared birthday) =

P(shared birthday) = 1 – P(no shared birthday)  =

157

## Example 2

Suppose five people are in a room. What is the probability that there is at least one shared birthday among these five people?

Continuing the pattern of the previous example, the answer should be

$P$(shared birthday) =

Refer to section 4.4 Counting Techniques to rewrite this more compactly as

$P$(shared birthday) =

which makes it a bit easier to type into a calculator or computer, and which suggests a nice formula as we continue to expand the population of our group.

## Example 3

Suppose 30 people are in a room. What is the probability that there is at least one shared birthday among these 30 people?

$P$(shared birthday) =

## Example 4

Suppose there are $n$ people are in a room. What is the probability that there is at least one shared birthday among these $n$ people?

$P$(shared birthday) =

1       Suppose three people are in a room. What is the probability that there is at least one shared birthday among these three people?

2       Suppose thirty-three people are in a room. What is the probability that there is at least one shared birthday among these thirty-three people?

3       Suppose $n$ people are in a room. What is the probability that there is at least one shared birthday among these $n$ people?

## Probability using Permutations and Combinations

We can use permutations and combinations to help us answer more complex probability questions

This project requires the concepts of section 4.4 Counting Techniques.

---

**Permutation Formula**
Picking $r$ objects from $n$ total objects when order matters
$$_nP_r = \frac{n!}{(n-r)!}$$

---

**Combination Formula**
Picking $r$ objects from $n$ total objects when order doesn't matter
$$_nC_r = \frac{n!}{r!(n-r)!}$$

---

Example 1

A 4 digit PIN number is selected. What is the probability that there are no repeated digits?

There are 10 possible values for each digit of the PIN (namely: 0, 1, 2, 3, 4, 5, 6, 7, 8, 9), so there are $10 \cdot 10 \cdot 10 \cdot 10 = 10^4 = 10000$ total possible PIN numbers.

To have no repeated digits, all four digits would have to be different, which is selecting without replacement. We could either compute $10 \cdot 9 \cdot 8 \cdot 7$, or notice that this is the same as:

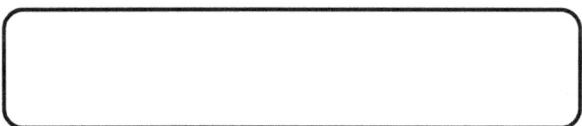

The probability of no repeated digits is the number of 4 digit PIN numbers with no repeated digits divided by the total number of 4 digit PIN numbers.

This probability is:

## Example 2

In a certain state's lottery, 48 balls numbered 1 through 48 are placed in a machine and six of them are drawn at random. If the six numbers drawn match the numbers that a player had chosen, the player wins $1,000,000. In this lottery, the order the numbers are drawn in doesn't matter. Compute the probability that you win the million-dollar prize if you purchase a single lottery ticket.

In order to compute the probability, we need to count the total number of ways six numbers can be drawn, and the number of ways the six numbers on the player's ticket could match the six numbers drawn from the machine. Since there is no stipulation that the numbers be in any particular order, the number of possible outcomes of the lottery drawing is

[ ]. Of these possible outcomes, only one would match all six numbers on the player's ticket, so the probability of winning the grand prize is:

[ ]

## Example 3

In the state lottery from the previous example, if five of the six numbers drawn match the numbers that a player has chosen, the player wins a second prize of $1,000. Compute the probability that you win the second prize if you purchase a single lottery ticket.

As above, the number of possible outcomes of the lottery drawing is $_{48}C_6 = 12,271,512$. In order to win the second prize, five of the six numbers on the ticket must match five of the six winning numbers; in other words, we must have chosen five of the six winning numbers and one of the 42 losing numbers. The number of ways to choose 5 out of the 6 winning numbers is given by $_6C_5 = 6$ and the number of ways to choose 1 out of the 42 losing numbers is given by $_{42}C_1 = 42$. Thus the number of favorable outcomes is then given by the Basic Counting Rule: $_6C_5 \cdot {_{42}C_1} = 6 \cdot 42 = 252$. So the probability of winning the second prize is.

$$\frac{\left(_6C_5\right)\left(_{42}C_1\right)}{_{48}C_6} = \left[ \quad \right]$$

## Example 4

Compute the probability of randomly drawing five cards from a deck and getting exactly one Ace.

In many card games (such as poker) the order in which the cards are drawn is not important (since the player may rearrange the cards in his hand any way she chooses); in the problems that follow, we will assume that this is the case unless otherwise stated. Thus we use combinations to compute the possible number of 5-card hands, $_{52}C_5$. This number will go in the denominator of our probability formula, since it is the number of possible outcomes.

For the numerator, we need the number of ways to draw one Ace and four other cards (none of them Aces) from the deck. Since there are four Aces and we want exactly one of them, there will be $_4C_1$ ways to select one Ace; since there are 48 non-Aces and we want 4 of them, there will be $_{48}C_4$ ways to select the four non-Aces. Now we use the Basic Counting Rule to calculate that there will be $_4C_1 \cdot {}_{48}C_4$ ways to choose one ace and four non-Aces.

Putting this all together, we have

$P(\text{one Ace}) =$

## Example 5

Compute the probability of randomly drawing five cards from a deck and getting exactly two Aces.

The solution is similar to the previous example, except now we are choosing 2 Aces out of 4 and 3 non-Aces out of 48:

## Example 6

Compute the probability of randomly drawing five cards from a deck and getting exactly three Hearts.

The solution is similar to the previous example, except now we are choosing three Hearts out of 13 and two non-Hearts out of 39:

---

**Permutation Formula**
Picking $r$ objects from $n$ total objects when order matters

$$_nP_r = \frac{n!}{(n-r)!}$$

---

**Combination Formula**
Picking $r$ objects from $n$ total objects when order doesn't matter

$$_nC_r = \frac{n!}{r!(n-r)!}$$

---

**1**    A 4 letter Code word is selected. What is the probability that there are no repeated letters?

**2**    Compute the probability of randomly drawing five cards from a deck and getting exactly two Spades.

1    In a certain state's lottery, 25 balls numbered 1 through 25 are placed in a machine and six of them are drawn at random. If the six numbers drawn match the numbers that a player had chosen, the player wins $30,000. In this lottery, the order the numbers are drawn in doesn't matter. Compute the probability that you win the prize if you purchase a single lottery ticket.

2    In another state's lottery, 25 balls numbered 1 through 25 are placed in a machine and six of them are drawn at random. If the six numbers drawn match the numbers that a player had chosen, the player wins $100,000. In this lottery, the order the numbers are drawn in does matter. Compute the probability that you win the prize if you purchase a single lottery ticket.

As part of a study, you agree to be tested for a rare and debilitating disease that affects 0.1% of the population. You are told that 5% of the population have indicators that cause the test to return positive even though they do not have the disease. Also, the test is positive for everyone who has the disease. A few days later you receive a call saying that you tested positive for the disease.

| "A rare and debilitating disease affects 0.1% of the population." |
| --- |
| $P(D)=$ |
| "The test is positive for everyone who has the disease." |
| $P(T|D)=$ |
| "5% of the population have indicators that cause the test to return positive even though they do not have the disease." |
| $P(T|not\,D)=$ |

Complete the table starting with a population of 100,000:

|  | $T$ | *not T* |  |
| --- | --- | --- | --- |
| $D$ |  |  |  |
| *not D* |  |  |  |
|  |  |  | 100,000 |

What is the probability that you have the disease?

$P(D|T)=$

## Derive the simple form of Bayes' Probability Formula

Suppose that $P(A)$, $P(B)$, and $P(B|A)$ are known.

Use the fact that $P(A) \cdot P(B|A) = P(A \text{ and } B)$ and the fact that $P(B|A)$ is known to show $P(A \text{ and } B)$:

| | B | not B | |
|---|---|---|---|
| A | | | $P(A)$ |
| not A | | | |
| | $P(B)$ | | 1 |

Use this to write the formula for $P(A|B)$:

We use this formula when both that $P(A)$ and $P(B)$ are known.

## Derive Bayes' Probability Formula

Suppose that $P(A)$, $P(B|A)$ and $P(B|not\ A)$ are known. Since $P(not\ A)=1-P(A)$, we can also consider $P(not\ A)$ to be known.

Use the fact that $P(A)\cdot P(B|A)=P(A\ and\ B)$ and the fact that both $P(B|A)$ and $P(B|not\ A)$ are known to show $P(A\ and\ B)$ and $P(not\ A\ and\ B)$.

Use the table to compute $P(B)$.

|        | B |     | not B |        |
|--------|---|-----|-------|--------|
| A      |   |     |       | $P(A)$ |
| not A  |   |     |       | $P(not\ A)$ |
|        |   |     |       | 1      |

Use this to write the formula for $P(A|B)$:

<u>Bayes Probability</u> allows you to change the perspective of the information you have. That is, if you know $P(A|B)$ you can find $P(B|A)$.

Previously, we knew $P(\text{positive test} | \text{not have the disease})$ and wanted to learn $P(\text{not have the disease} | \text{positive test})$.

**Ex 1:** Before eliminating smoking on campus, the administration studied the change. They wanted to know what percentage of women are smokers. According to the Centers for Disease Control, CDC, 14% of Californians smoke and 37% of smokers are women, what is percentage of women who are smokers? Assume that 52% of the population are women.

Again, we are changing the perspective. We know $P(\text{woman} | \text{smoker})$ and want to learn
$P(\text{smoker} | \text{woman})$.

State what we know:

$P(W) =$ $\qquad$ $P(S) =$ $\qquad$ $P(W | S) =$

| | W | not W | |
|---|---|---|---|
| S | | | |
| not S | | | |
| | | | 30,000 |

**Ex 2:** (From *The Language of Mathematics* by Keith Devlin page 289) A certain town has two taxi companies, Blue Cabs and Black Cabs. Blue Cabs has 15 taxis and Black Cabs has 85 taxis. Late one night, there is a hit-and-run involving a taxi. All the town's 100 taxis are on the streets at the time of the accident. A witness sees the collision and claims that a blue taxi was involved. At the request of the police, the witness undergoes a vision test under conditions similar to those on the night in question. Presented repeatedly with a blue taxi and a black taxi, in random order, he shows he can successfully identify the color of the taxi four times out of five. If you were investigating the case, which taxi company would you think is most likely to have been involved in the collision?

Let B be a collision with a cab from Blue Taxi company. Let b claims to have seen a cab of the Blue Taxi company.

|  | *b* | *not b* |  |
|---|---|---|---|
| B |  |  |  |
| not B |  |  |  |
|  |  |  | 1 |

As part of a study, you agree to be tested for a rare and debilitating disease that affects 1% of the population. You are told that 15% of the population have indicators that cause the test to return positive even though they do not have the disease. Also, the test is positive for everyone who has the disease. A few days later you receive a call saying that you tested positive for the disease.

| |
|---|
| "A rare and debilitating disease affects 1% of the population." |
| |
| "The test is positive for everyone who has the disease." |
| |
| "15% of the population have indicators that cause the test to return positive even though they do not have the disease." |
| |

What is the probability that you have the disease?

As part of a study, you agree to be tested for a rare and debilitating disease that affects 0.5% of the population. You are told that 10% of the population have indicators that cause the test to return positive even though they do not have the disease. Also, the test is positive for 90% of those who have the disease. A few days later you receive a call saying that you tested positive for the disease.

**1**      Translate. (Use D for having the disease and T for testing positive.)

| "A rare and debilitating disease affects 0.5% of the population." |
|---|
|  |
| "10% of the population have indicators that cause the test to return positive even though they do not have the disease." |
|  |
| "The test is positive for 90% of those who have the disease." |
|  |

**2**      Complete the table for 100,000 people

|  | T | Not T |  |
|---|---|---|---|
| D |  |  |  |
| Not D |  |  |  |
|  |  |  | 100,000 |

**3**      What is the probability that you have the disease?

Use $P(A)$, $P(B)$, and $P(B|A)$ to derive the formula for $P(A|B)$.

**1**    Use the table to derive the formula:

|        | B | Not B |   |
|--------|---|-------|---|
| A      |   |       |   |
| Not A  |   |       |   |
|        |   |       | 1 |

**2**    Use the identity to derive the formula:

$$P(A \text{ and } B) = P(B \text{ and } A)$$

Use $P(A)$, $P(B|A)$ and $P(B|\textit{not } A)$ to derive the formula for $P(A|B)$.

**3**    Since _____, we can also use $P(\textit{not } A)$.

**4**    Use the table to derive the formula:

|  |  | B | Not B |  |
|---|---|---|---|---|
| A |  |  |  |  |
| Not A |  |  |  |  |
|  |  |  |  | 1 |

1       Before eliminating subsidized low-income automobile insurance, the governor wanted to study the effect on public safety. According to the NHTSA ,15% of drivers are reckless. According to the Commission on State Finance, 30% of drivers receive insurance subsidies. According to the CHP, 45% of reckless driving citations are given to drivers who receive insurance subsidies. What percentage of low income drivers are reckless?

**1**     A certain town has two taxi companies, Blue Cabs and Black Cabs. Blue Cabs has 15 taxis and Black Cabs has 85 taxis. Late one night, there is a hit-and-run involving a taxi. All the town's 100 taxis are on the streets at the time of the accident. A witness sees the collision and claims that a blue taxi was involved. At the request of the police, the witness undergoes a vision test under conditions similar to those on the night in question. Presented repeatedly with a blue taxi and a black taxi, in random order, he shows he can successfully identify the color of the taxi four times out of five. If you were investigating the case, which taxi company would you think is most likely to have been involved in the collision?

Let B be a collision with a cab from Blue Taxi company. Let b claims to have seen a cab of the Blue Taxi company.

|       | b | not b |   |
|-------|---|-------|---|
| B     |   |       |   |
| not B |   |       |   |
|       |   |       | 1 |

Mean: $\bar{x} = \dfrac{\sum x}{n}$, where $n$ is the number of $x$s and $\sum x$ is the sum of the $x$s.

Median: The middle value when the data is arranged in order.

Mode: The mode is the (those) value(s) which occur most often.

Mean: From a frequency distribution: $\bar{x} = \dfrac{\sum (x \cdot f)}{\sum f}$, where $f$ is the frequency of $x$.

Weighted Mean: $\bar{x} = \dfrac{\sum (x \cdot f)}{\sum f}$, where $f$ is the weight of $x$.

Range: (Largest Value) - (Smallest Value)

## Example 1

Find the mean, mode and median of the following numbers:

8, 9, 9, 9, 12, 13, 14, 21, 28, 37, 56, 58, 67, 69, 73

The mean is:

The mode is:

The median is:

## Example 2

Given the frequency distribution for the following find the mean and mode.

| Value: | Frequency: | |
|--------|-----------|---|
| 6 | 2 | |
| 10 | 3 | |
| 14 | 8 | |
| 18 | 2 | |
| 22 | 1 | |
| Sum: | | |

The mean is:

The mode is:

$$s = \sqrt{\frac{\sum (x - \bar{x})^2}{n-1}} \qquad\qquad \bar{x} = \frac{\sum x}{n}$$

## Example 3
Find the Standard Deviation of the following data:

| Data | 10 | 10 | 10 | |
|------|----|----|----|------|
| Deviation | | | | sum: |
| Deviation Squared | | | | |

$$s = \sqrt{\rule{3cm}{0pt}} = \sqrt{\rule{2cm}{0pt}}$$

## Example 4

| Data | 5 | 10 | 15 | |
|------|---|----|----|------|
| Deviation | | | | sum: |
| Deviation Squared | | | | |

$$s = \sqrt{\rule{3cm}{0pt}} = \sqrt{\rule{2cm}{0pt}}$$

**Example 5**

| Data | 2 | 7 | 21 | |
|------|---|---|-----|-------|
| Deviation | | | | sum: |
| Deviation Squared | | | | |

$$s = \sqrt{\underline{\hspace{3cm}}} = \sqrt{\underline{\hspace{2cm}}}$$

**Example 6**

| Data | 1 | 3 | 8 | 4 | 6 | 4 | 3 | 9 | 7 | |
|------|---|---|---|---|---|---|---|---|---|------|
| Deviation | | | | | | | | | | sum: |
| Deviation Squared | | | | | | | | | | |

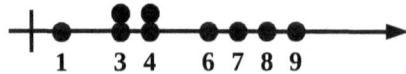

Mean: $\bar{x} =$

Standard Deviation: $s = \sqrt{\underline{\hspace{3cm}}} = \sqrt{\underline{\hspace{2cm}}}$

## Draw One Card

| | |
|---|---|
| P(Heart) = | P(5) = |
| P(Queen) = | P(Club) = |

| | |
|---|---|
| P(Club or Heart) = | P(Jack or 8) = |
| P(Club and Heart) = | P(Jack and 8) = |
| P(Black) = | P(Even) = |
| When two events are mutually exclusive: | |
| P(A or B) = | P(A and B) = |

| | |
|---|---|
| P(Club or 6) = | P(Jack or Spade) = |
| P(Club and 6) = | P(Jack and Spade) = |
| P(Diamond and Ace) = | P(Even and Black) = |
| When two events are not mutually exclusive: | |
| P(A or B) = | P(A and B) is not zero. |

Conditional Probability from a Table

Suppose that the outcome is already known to have a <u>given</u> characteristic.

Definition: $P(A|B)=\dfrac{n(A \text{ and } B)}{n(B)}$. $P(A|B)$ is read "The Probability of A given B."

|  | California | Nevada | total: |
|---|---|---|---|
| Senators: | 2 | 2 | 4 |
| Congressional Representatives: | 53 | 3 | 56 |
| total: | 55 | 5 | 60 |

Supposed you selected one of the politicians at random. Now, supposes you are given that the person is a Nevadan. "<u>Given</u> that a Nevadan was selected what is the probability that the Nevadan is a Senator?" To answer this we only consider the Nevadans. There are 5 Nevadans, 2 senators and 3 representatives. We can write $P$(Senator | Nevadan). This is read, "the probability of selecting a senator given a Nevadan."

$$P(\text{senator}|\text{Nevadan})=\frac{n(\text{senator and Nevadan})}{n(\text{Nevadan})}$$

What is the probability that if a Nevadan is selected, a senator was selected?

|  | California | Nevada | total: |
|---|---|---|---|
| Senators: | 2 | 2 | 4 |
| Congressional Representatives: | 53 | 3 | 56 |
| total: | 55 | 5 | 60 |

Find the following probabilities from the table:

$P$(California) =                     $P$(Nevadan) =

$P$(Californian or Senator) =

$P$(Nevadan or Representative) =

$P$(Nevadan and Senator) =                     $P$(Californian and Representative) =

$P$(Nevadan | Senator) =                     $P$(Californian | Senator) =

$P$(Senator | Nevadan) =                     $P$(Senator | Californian) =

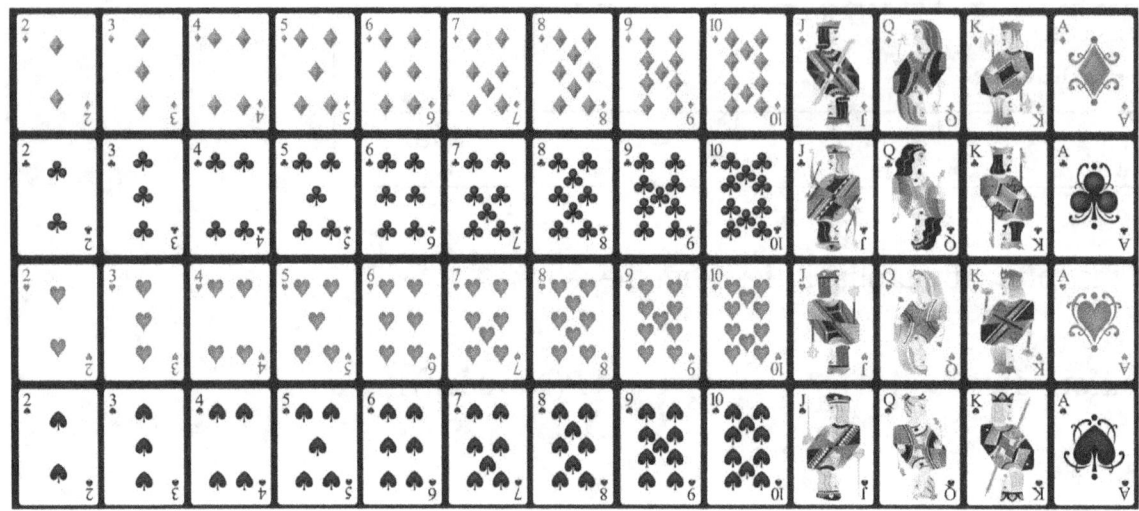

### Draw Two Cards with Replacement

| P(Ace and 9) = | P(Diamond and Club) = |
|---|---|
| P(Club and Heart) = | P(Jack and 8) = |
| P(Ace and Ace) = | P(Diamond and Diamond) = |
| When two events are independent: | |
| P(A and B) = | |

### Draw Two Cards without Replacement

| P(Ace and 9) = | P(Diamond and Club) = |
|---|---|
| P(Club and Heart) = | P(Jack and 8) = |
| P(Ace and Ace) = | P(Diamond and Diamond) = |
| When two events are not independent: | |
| P(A and B) = | |

Probability Rules

If two events are **mutually exclusive**, then P(A or B) =_____

and P(A and B) = ___.

If two events are **not mutually exclusive**, then P(A or B)

=_____

and P(A and B) = ___.

If two events are **independent**, then P(A and B) =_____

and P(A | B) = _____, and P(B | A) = _____.

If two events are **not independent**, then P(A and B) =_____ or

_____,

P(A | B) = _____, and P(B | A) = _____.

1    Find the probability that a randomly chosen California has a college degree and
     is a Sacramentan given that: 30% of Californians have college degrees, only
     9% of Californians are from Sacramento, and that 35% of Sacramentans have a
     college degree.

2   Given two events A and B that are not independent such that $P(B \mid A) = 0.45$, and $P(A \text{ and } B) = 0.16$, find $P(A)$.

3   Given two events A and B that are not independent such that $P(\text{not } A \mid B) = 0.59$, and $P(\text{not } B) = 0.27$, find $P(A \text{ and } B)$.

4   Given that $P(A \text{ and } B) = 0.16$, $P(A \text{ or } B) = 0.58$ and $P(B) = 0.3$, find $P(A)$.

5   Given two events A and B that are not independent such that $P(A) = 0.32$, $P(A \text{ or } B) = 0.61$, and $P(A \text{ and } B) = 0.12$, find $P(A \mid B)$. (Hint: Find $P(B)$ first.)

1      Choosing what to wear you can choose from 3 pairs of shoes, 8 shirts, and 2 jeans. How many different outfits can you make?

2      All of the license plates in a particular state feature a number followed by two letters followed by two digits (e.g. 1AB12). How many different license plate numbers are available?

Permutations: $P(n,r) = \dfrac{n!}{(n-r)!}$          Combinations: $C(n,r) = \dfrac{n!}{r!(n-r)!}$

where, $n! = n \cdot (n-1) \cdot (n-2) \cdots 3 \cdot 2 \cdot 1$

Order: Both Combinations and Permutations count collections of $r$ things or people

selected from among $n$ things or people. The difference is how those

collections are considered. If order matters within the collection of $r$ things or

people, then we count the number of these collections, called permutations,

with $P(n,r)$. Otherwise, when there is no order within the collections, we

count the number of these collections, called combinations, with $C(n,r)$.

3      In how many ways can a president, vice president and treasurer be elected at a school with 210 students?

**4**   In how many ways can a committee of three students be formed at a school with 210 students?

**5**   The theatre director must chose five plays to present in a season. There are props and costumes for 19 different plays available. How many different seasons are there?

**6**   The once five plays are chosen, how many different orders are there to present the plays?

| Permutations: | Combinations: |
|:---:|:---:|
| Choosing Passwords | Choosing "Lotto" numbers |
| Choosing People for Different Offices | Choosing People for Committees |
| Choosing ZIP codes to visit in order | Choosing ZIP codes to send surveys to |
| Choosing who will receive (different) presents | Choosing a greeting card list |
| Scheduling batting order | Choosing who will go to bat |
| Ordering pianists in a recital | Choosing pianists to play in a recital |
| *Arranging* different things | *Arranging* identical things |

Example of "Arrangements":

**7** How many 8-digit binary numbers (base-2) have exactly 3 "1s"?

This is a question of arrangements. ⊔⊔⊔⊔⊔⊔⊔⊔
1 2 3 4 5 6 7 8

The position numbers {1, 2, 3, 4, 5, 6, 7, 8} is the collection from which we must select three locations to place "1s." For example, if we chose {1, 2, 3} we place "1s" there and "0s" elsewhere. This give us: 11100000 Similarly, if we chose {2,1,3} we also place "1s" in the first three positions. Again, we have: 11100000.

**8** How many ways can it rain three times in a week?

**9** How many ways can 3 deluxe model homes and 5 standard model homes be arranged?

**10** On your pet's birthday week, you have three surprise gifts, a bone, a new collar and a ball. How many ways are there of giving your pet these surprise gifts, with no more that one gift given on each day?

⊔⊔⊔⊔⊔⊔⊔
M T W R F S U

Consider seven coin tosses. Complete the table by asking yourself, "In how many ways can you toss exactly $x$ tails?"

| exactly $x$ tails: | Number of ways: $n(x)$ |
|---|---|
| 0 | $C(7,0)=$ |
| 1 | $C(7,1)=$ |
| 2 | $C(7,2)=$ |
| 3 | $C(7,3)=$ |
| 4 | $C(7,4)=$ |
| 5 | $C(7,5)=$ |
| 6 | $C(7,6)=$ |
| 7 | $C(7,7)=$ |
| Total: | |

**11** In how many ways can you toss a coin seven times and have three tails?

**12** In how many ways can you toss a coin seven times and have three or four tails?

Complements Principle of Counting:

$$\begin{pmatrix} \text{Number of ways} \\ \text{event A} \\ \text{does not occur} \end{pmatrix} = \begin{pmatrix} \text{Total number of} \\ \text{events that can occurs} \end{pmatrix} - \begin{pmatrix} \text{Number of ways} \\ \text{event A occurs} \end{pmatrix}$$

**13** In how many ways can you toss a coin seven times and have less than seven tails?

**14** In how many ways can you toss a coin fifteen times and have at most fourteen tails?

Binomial Probability: $p(x) = C(n,x) p^x q^{n-x}$, where there are $n$ independent trails, $x$ successes and the probability of a success in one trail is $p$. By the complement rule $q = 1 - p$.

**Ex 1:** An orange tree has a 65% chance of surviving a frost in an orchard with 6 trees.

| Number of trees which survive: $x$ | $p(x)$ | $x \cdot p(x)$ |
|---|---|---|
| 0 | | |
| 1 | | |
| 2 | | |
| 3 | | |
| 4 | | |
| 5 | | |
| 6 | | |
| | Sum: $\sum (x \cdot p(x))$ | |

Referring to the table, what is the probability that:

**Ex 2:** Exactly 4 orange trees survive?

**Ex 3:** At least half the orchard survives?

**Ex 4:** How many trees in the orchard do we expect to survive?

Expected Value: Given a random variable $x$ ranging from $x_1$ through $x_n$, and the corresponding probability of each through $p(x_n)$, the expected value is

$$x_1 \cdot p(x_1) + x_2 \cdot p(x_2) + \cdots + x_n \cdot p(x_n).$$

**Ex 1:** A certain game of chance involves a 6-sided dice. It costs \$1 to play. If the dice lands showing a 3 or 4, then \$2 is won. If the dice lands on a 5, then \$5 is won. Find the expected value for this game. When you have found the expected value, answer the question "is this a fair game"?

| $x$ | $p(x)$ | $x \cdot p(x)$ |
|---|---|---|
|  |  |  |
|  |  |  |
|  |  |  |
|  | Expected Value: |  |

**Ex 2:** A certain firm has three salary echelons. There are 15 employees. 6 earn $32,000 per year, 5 earn $34,000 per year and 4 earn $36,000 per year. If an employee is chosen at random find that person's expected salary.

| $x$ | $p(x)$ | $x \cdot p(x)$ |
|---|---|---|
| | | |
| | | |
| | | |
| | Expected Value: | |

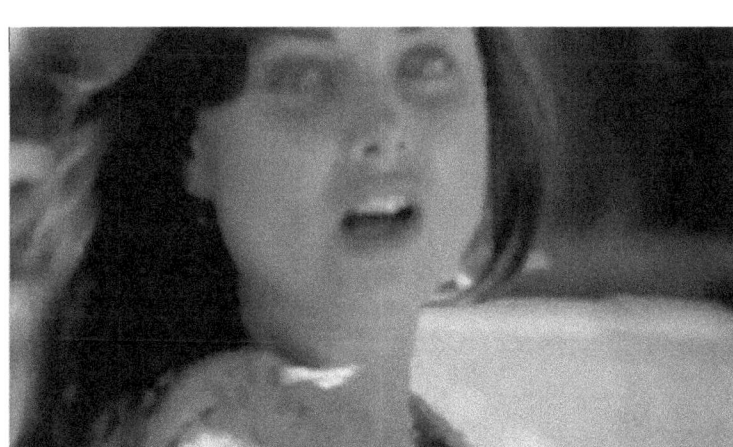

• How far was the young woman thrown?

_____ m or _____ ft

• What injuries did she suffer?



• What was the speed limit?

_____ km/h or _____ mph

• How fast was he going when he hit her?

_____ km/h or _____ mph

• The officer knows this because he analyzed the _____ and these

_____.

• How fast was he going when he first saw the young women?

_____ km/h or _____ mph

*Let's change just one thing...*

• How fast is he going now when he first sees the young women?

_____ km/h or _____ mph

• This time he hits her at what speed?

_____ km/h or _____ mph

• What injuries would she have suffered?

• The injuries suffered by the young woman were influenced by what two speeds?

1)

2)

• What is the formula for kinetic energy?

• This formula determines what two ways that energy was dissipated in the collision?

1)

2)

• In the speeding collision, the young woman was hit at _____ mph. By going _____ mph slower, the second collision happened at only _____ mph. What part of the formula for kinetic energy determines that such a small change in speed has such a dramatic effect on the collision?

## Translate

"Varies" means "is a multiple of."

$$\underbrace{\text{varies}}$$
$$\underbrace{\text{is}} \quad \underbrace{\text{a multiple}} \quad \underbrace{\text{of}}$$
$$\underline{\hspace{2cm}} \quad \underline{\hspace{1.5cm}} \quad \underline{\hspace{1.5cm}}$$

"Directly" is the numerator.

"Inversely" and "indirectly" is the denominator.

"Jointly" means "the product of."

1) Translate to get the General Equation

2) Substitute the Given Information

3) Solve for k

4) Find the Specific Equation

5) Substitute the Question and Solve

6) State the Answer with Units

---

Translate: The amount of energy in a collision varies directly with the mass and the square of the velocity. (Use "$E$" for energy, "$m$" for the mass, and "$v$" for the velocity.)

---

Translate: The amount of energy in joules released in a collision is ½ the mass in kilograms times the square of the velocity in meters per second.

**1**   The collision impact of an automobile varies jointly with its weight and the square of its speed. Suppose a 2000-lb car traveling at 55 mph has a collision impact of 6.1. What is the collision impact of a 3000-lb SUV traveling car at 65 mph?

**2**   How many Joules of energy was released when the driver of a commercial truck loses control and collides with a house? The truck weights 11,500 lbs (4535 kg) and was traveling at 40 mph (18 m/s).

**3**   How many times larger is 15 ft than 5 ft?

**4**   How many times larger is 87 hours than 3 hours?

**5**    How many times more energy was released in a collision at 50 mph than at 35 mph?

**6**    How many times more energy was released in a collision at 40 mph than at 15 mph?

**7**    A pedestrian hit by the driver of an automobile at 25 mph has an 80% chance of surviving the collision. At 35 mph the change of being killed as a result of the collision is 80%. How many times more energy was released in a collision at 35 mph than at 25 mph?

**8**    On a fenced-off track an engineer applies the brakes to skid to a stop from a speed of 35 mph. The skid marks are 93 ft long. How long would the skid marks have been had the engineer began to skid at 49 mph?

1      The injuries suffered by the young woman were influenced by what two speeds?

_____

_____

2      What is the formula for kinetic energy in Joules?

_____

3      How many Joules of energy was released when the driver of an SUV loses control and collides with a parked car? The SUV weighs 8,250 lbs (3742 kg) and was traveling at 45 mph (20 m/s).

4      On a fenced-off track an engineer applies the brakes to skid to a stop from a speed of 50 mph. The skid marks are 128 ft long. How long would the skid marks have been had the engineer began to skid at 35 mph?

1　　　When the driver hit the women causing severe brain damage, what was he doing?

_____

2　　　What part of the formula for kinetic energy determines that such a small change in speed has such a dramatic effect on the collision?

_____

3　　　The collision impact of an automobile varies jointly with its weight and the square of its speed. Suppose a 2500-lb car traveling at 35 mph has a collision impact of 3.1. What is the collision impact of a 3000-lb SUV traveling car at 50 mph?

4　　　How many times more energy was released in a collision at 50 mph than at 35 mph?

1      What is the lesson of the public safety video?

_____

_____

2      A CHP officer measures the skid marks leading to a collision. The skid marks are 110 ft long. An experiment with the same type of car tells the officer that a car traveling at the speed limit would leave skids 45 ft long. The speed limit is 35 mph. How fast was the driver traveling?

3      A pedestrian hit by the driver of an automobile at 25 mph has an 80% chance of surviving the collision. At 35 mph the change of being killed as a result of the collision is 80%. How many times more energy was released in a collision at 40 mph than at 30 mph?

Recall that: $\text{base} \times \text{percent} = \text{amount}$.

Ex      If 10% of the people at the airport are elderly and there are 270 people, how many are elderly?

Ex      A statistician surveys 400 people on the bus and 600 people at the airport. Only 10% of those who fly are elderly, while 20% of those on the bus are elderly. The sample is 14% elderly. Complete the table.

| | base | percentage | amount |
|---|---|---|---|
| Bus Riders | | | |
| Airport | | | |
| Sample | | | |

Ex A statistician needs a sample population of 1000 people that is 18% elderly. On the bus 20% are elderly. At an airport 10% are elderly. How many of each population should be surveyed?

| WORK | base | percentage | amount |
|---|---|---|---|
| Bus Riders | | | |
| Airport | | | |
| Sample | | | |

| CHECK | base | percentage | amount |
|---|---|---|---|
| Bus Riders | | | |
| Airport | | | |
| Sample | | | |

1      A statistician needs a sample population of 1500 people that is 12% elderly. On
       the bus 20% are elderly. At an airport 10% are elderly. How many of each
       population should be surveyed?

| WORK | base | percentage | amount |
|------|------|------------|--------|
| Bus Riders | | | |
| Airport | | | |
| Sample | | | |

| CHECK | base | percentage | amount |
|-------|------|------------|--------|
| Bus Riders | | | |
| Airport | | | |
| Sample | | | |

1    A statistician needs a sample population of 700 people that is 18% elderly. On the bus 20% are elderly. At an airport 10% are elderly. How many of each population should be surveyed?

| WORK | base | percentage | amount |
|------|------|------------|--------|
| Bus Riders | | | |
| Airport | | | |
| Sample | | | |

| CHECK | base | percentage | amount |
|-------|------|------------|--------|
| Bus Riders | | | |
| Airport | | | |
| Sample | | | |

1      A statistician needs a sample population of 2400 people that is 13% elderly. On the bus 20% are elderly. At an airport 10% are elderly. How many of each population should be surveyed?

| WORK | base | percentage | amount |
|---|---|---|---|
| Bus Riders | | | |
| Airport | | | |
| Sample | | | |

| CHECK | base | percentage | amount |
|---|---|---|---|
| Bus Riders | | | |
| Airport | | | |
| Sample | | | |

1    A statistician needs a sample population of 1200 people that is 13% elderly. On the bus 15% are elderly. At an airport 7% are elderly. How many of each population should be surveyed?

| WORK | base | percentage | amount |
|---|---|---|---|
| Bus Riders | | | |
| Airport | | | |
| Sample | | | |

| CHECK | base | percentage | amount |
|---|---|---|---|
| Bus Riders | | | |
| Airport | | | |
| Sample | | | |

1    The revenue earned often increases then decreases as price increases. The revenue in thousands of dollars is a function of the price in dollars: $R = -8 p^2 + 128 p$. Find the price that maximizes the revenue and the maximum revenue.

$_____$ is the price that maximizes the revenue.

$_____$ thousand is the maximum revenue.

2    The revenue in thousands of dollars is a function of the price in dollars: $R = -22 p^2 + 132 p$. Find the price that maximizes the revenue and the maximum revenue.

$_____$ is the price that maximizes the revenue.

$_____$ thousand is the maximum revenue.

1    The revenue in thousands of dollars is a function of the price in dollars:

$R = -17 p^2 + 442 p$. Find the price that maximizes the revenue and the maximum

revenue.

$ _____ is the price that maximizes the revenue.

$ _____ thousand is the maximum revenue.

1    The revenue in thousands of dollars is a function of the price in dollars:

$R = -25p^2 + 650p$. Find the price that maximizes the revenue and the maximum

revenue.

$_____ is the price that maximizes the revenue.

$_____ thousand is the maximum revenue.

1     The revenue in thousands of dollars is a function of the price in dollars:

$R = -17p^2 + 646p$. Find the price that maximizes the revenue and the maximum

revenue.

$_____ is the price that maximizes the revenue.

$_____ thousand is the maximum revenue.

1    The revenue in thousands of dollars is a function of the price in dollars: $R=-4p^2+216p$. Find the price that maximizes the revenue and the maximum revenue.

$_____ is the price that maximizes the revenue.

$_____ thousand is the maximum revenue.

1    A ball was kicked from a height of 3.3 feet. It was in flight for 1.8 seconds.

The initial vertical velocity of the ball, $v_0$, is _____ feet per second. (Round your velocity to the nearest hundredth. Use this rounded number for the following question.)

The time at which the maximum height is reached, $t_{vertex}$, is _____ seconds. (Round your time to the nearest hundredth. Use this rounded number for the following question.)

The maximum height reached, $h(t_{vertex})$, is _____ feet. (Round your height to the nearest hundredth.)

**2**     A ball was kicked from a height of 2.5 feet. It was in flight for 2.8 seconds.

The initial vertical velocity of the ball, $v_0$, is _____ feet per second. (Round your velocity to the nearest hundredth. Use this rounded number for the following question.)

The time at which the maximum height is reached, $t_{vertex}$, is _____ seconds. (Round your time to the nearest hundredth. Use this rounded number for the following question.)

The maximum height reached, $h(t_{vertex})$, is _____ feet. (Round your height to the nearest hundredth.)

1    A ball was kicked from a height of 2.3 feet. It was in flight for 4.5 seconds.

The initial vertical velocity of the ball, $v_0$, is _____ feet per second. (Round your velocity to the nearest hundredth. Use this rounded number for the following question.)

The time at which the maximum height is reached, $t_{vertex}$, is _____ seconds. (Round your time to the nearest hundredth. Use this rounded number for the following question.)

The maximum height reached, $h(t_{vertex})$, is _____ feet. (Round your height to the nearest hundredth.)

1    A ball was kicked from a height of 3 feet. It was in flight for 6 seconds.

The initial vertical velocity of the ball, $v_0$, is _____ feet per second. (Round your velocity to the nearest hundredth. Use this rounded number for the following question.)

The time at which the maximum height is reached, $t_{vertex}$, is _____ seconds. (Round your time to the nearest hundredth. Use this rounded number for the following question.)

The maximum height reached, $h(t_{vertex})$, is _____ feet. (Round your height to the nearest hundredth.)

1    A ball was launched from a catapult height of 17 feet. It was in flight for 7

seconds.

The initial vertical velocity of the ball, $v_0$, is _____ feet per second. (Round

your velocity to the nearest hundredth. Use this rounded number for the following

question.)

The time at which the maximum height is reached, $t_{vertex}$, is _____ seconds.

(Round your time to the nearest hundredth. Use this rounded number for the

following question.)

The maximum height reached, $h(t_{vertex})$, is _____ feet. (Round your height to

the nearest hundredth.)

The formula for accumulated interest, $A=P\left(1+r/n\right)^{nt}$, is an example of an exponential equation. The formula for continuous compounded interest, $A=Pe^{rt}$, is another example.

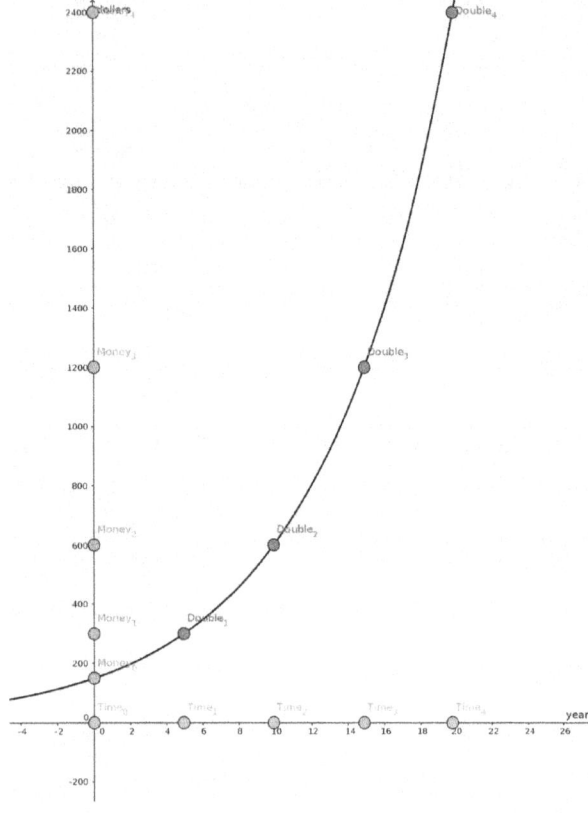

This is the graph of the amount of money in an account that earns interest at a constant rate of 15% per year starting with an initial investment of $150.

**The doubling time is constant.**

The doubling time for a quantity growing at a constant rate of $r$ per year is:

$T=\dfrac{\ln\left(2\right)}{r}$ , where $\ln\left(2\right)$ is the natural logarithm of 2.

Since $\ln\left(2\right)\approx0.693147181\approx0.69$ the rule can be approximated as:

$T\approx\dfrac{0.69}{r}$

For that reason, regardless of $P$ and $r$, all quantities that grow at a constant growth rate also have a constant doubling time.

Ex.: Graph the doubling function: $y=2^x$.

| $x$ | -5 | -4 | -3 | -2 | -1 | 0 | 1 | 2 | 3 | 4 | 5 |
|-----|----|----|----|----|----|---|---|---|---|---|---|
| $y=2^x$ | | | | | | | | | | | |

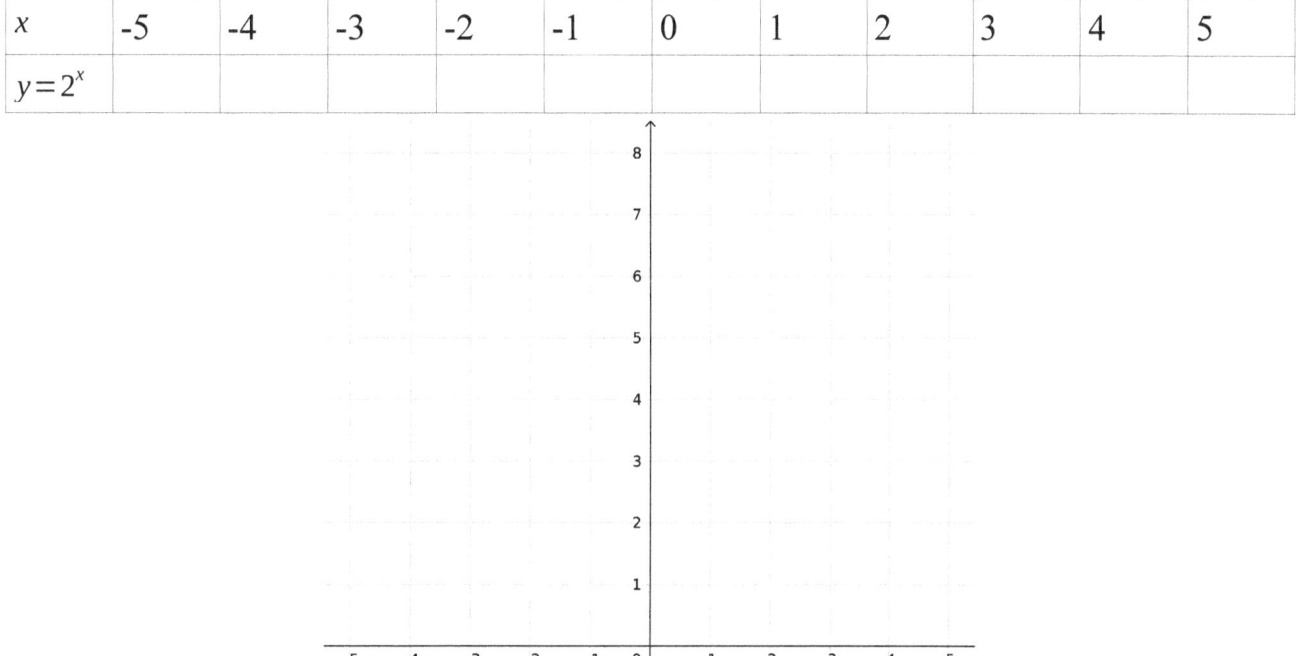

Compare an investment of $1000 that earns 5% interest compounded annually (Constant Growth Rate) with an account that grows by $50 each year (Constant Growth).

| | Constant Growth Rate (5%) | | | Constant Growth ($50) | | |
|---|---|---|---|---|---|---|
| Year | Starting Amt | Growth | Ending Amt | Starting Amt | Growth | Ending Amt |
| 1 | 1000.00 | 50.00 | 1050.00 | 1000.00 | 50.00 | 1050.00 |
| 2 | 1050.00 | 52.50 | 1102.50 | 1050.00 | 50.00 | 1100.00 |
| 3 | 1102.50 | 55.13 | 1157.63 | 1100.00 | 50.00 | 1150.00 |
| 4 | 1157.63 | 57.88 | 1215.51 | 1150.00 | 50.00 | 1200.00 |
| 5 | 1215.51 | 60.78 | 1276.28 | 1200.00 | 50.00 | 1250.00 |
| 6 | 1276.28 | 63.81 | 1340.10 | 1250.00 | 50.00 | 1300.00 |
| 7 | 1340.10 | 67.00 | 1407.10 | 1300.00 | 50.00 | 1350.00 |
| 8 | 1407.10 | 70.36 | 1477.46 | 1350.00 | 50.00 | 1400.00 |
| 9 | 1477.46 | 73.87 | 1551.33 | 1400.00 | 50.00 | 1450.00 |
| 10 | 1551.33 | 77.57 | 1628.89 | 1450.00 | 50.00 | 1500.00 |
| 11 | 1628.89 | 81.44 | 1710.34 | 1500.00 | 50.00 | 1550.00 |
| 12 | 1710.34 | 85.52 | 1795.86 | 1550.00 | 50.00 | 1600.00 |
| 13 | 1795.86 | 89.79 | 1885.65 | 1600.00 | 50.00 | 1650.00 |
| 14 | 1885.65 | 94.28 | 1979.93 | 1650.00 | 50.00 | 1700.00 |
| 15 | 1979.93 | 99.00 | 2078.93 | 1700.00 | 50.00 | 1750.00 |
| 16 | 2078.93 | 103.95 | 2182.87 | 1750.00 | 50.00 | 1800.00 |
| 17 | 2182.87 | 109.14 | 2292.02 | 1800.00 | 50.00 | 1850.00 |
| 18 | 2292.02 | 114.60 | 2406.62 | 1850.00 | 50.00 | 1900.00 |
| 19 | 2406.62 | 120.33 | 2526.95 | 1900.00 | 50.00 | 1950.00 |
| 20 | 2526.95 | 126.35 | 2653.30 | 1950.00 | 50.00 | 2000.00 |
| 21 | 2653.30 | 132.66 | 2785.96 | 2000.00 | 50.00 | 2050.00 |
| 22 | 2785.96 | 139.30 | 2925.26 | 2050.00 | 50.00 | 2100.00 |
| 23 | 2925.26 | 146.26 | 3071.52 | 2100.00 | 50.00 | 2150.00 |
| 24 | 3071.52 | 153.58 | 3225.10 | 2150.00 | 50.00 | 2200.00 |
| 25 | 3225.10 | 161.25 | 3386.35 | 2200.00 | 50.00 | 2250.00 |

We can graph the value of the account using the equation for compound interest: $A = P(1 + r/n)^{nt}$. In this case, $A = 1000(1 + 0.05/1)^{1t}$ or $A = 1000(1.05)^t$. Time, $t$, takes the place of $x$. The ending Amount, $A$, takes the place of $y$. The graph of the equation $y = 1000(1.05)^x$ is exponential.

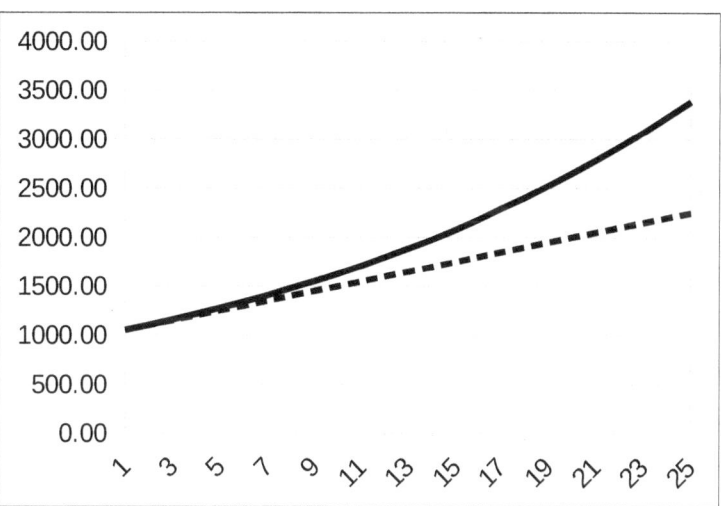

Ex.: Complete the table for the doubling time when $100 is invested in an account that earns 3.5% interest compounded annually.

| Value | $100 | | $200 | | $400 | | $800 | | $1600 | |
|-------|------|--|------|--|------|--|------|--|-------|--|
| Time  |      |  |      |  |      |  |      |  |       |  |

We let $x$ be the time, $t$, and let $y$ be the accumulated amount, $A$. With the graph we can see the value of the account increasing as time passes. (Beginning with $P = 100$, we are graphing the equation $y = 100(1.035)^x$.) Label the doubling time and the value of the account.

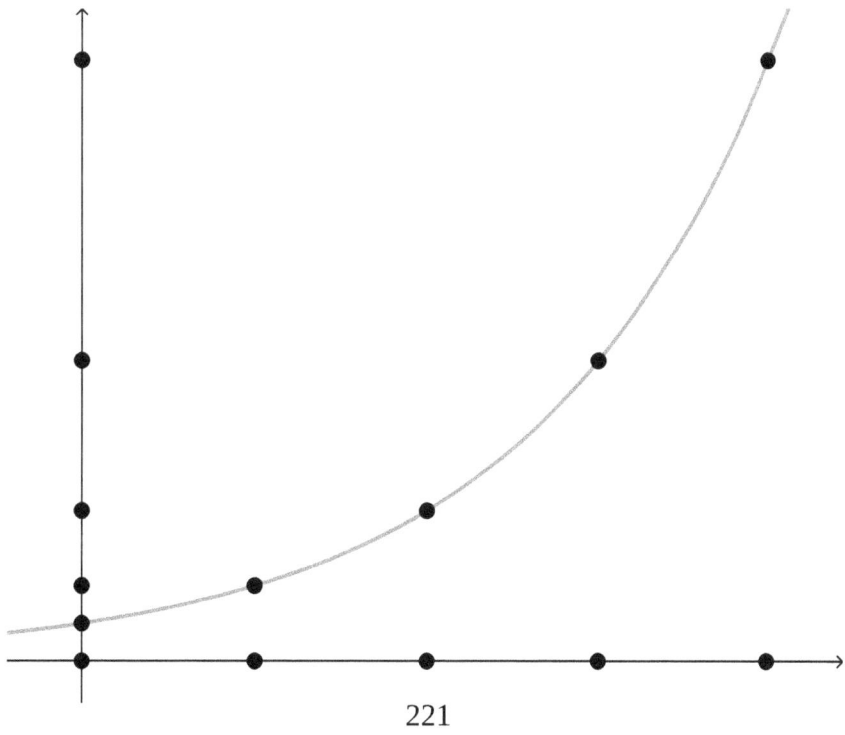

221

Ex.: Derive the formula for the doubling time given the growth rate, $r$.

We use the formula for continuous growth: $A=Pe^{rt}$. Since the doubling time is constant, for any beginning value, $P$, the ending value will be $A=2P$.

$T=\dfrac{\ln(2)}{r}$, where $T$ is the doubling time.

Ex.: As the pollution from automobiles grows at 7% per year, how long until the amount of pollution doubles?

Ex.: As the profits of a major corporation grows at 11% per year, how long until the profits double?

Ex.: Graph the population of a small town over time as its population increases from 4500 by 1% per year.

| x | | | | | |
|---|---|---|---|---|---|
| y | | | | | |

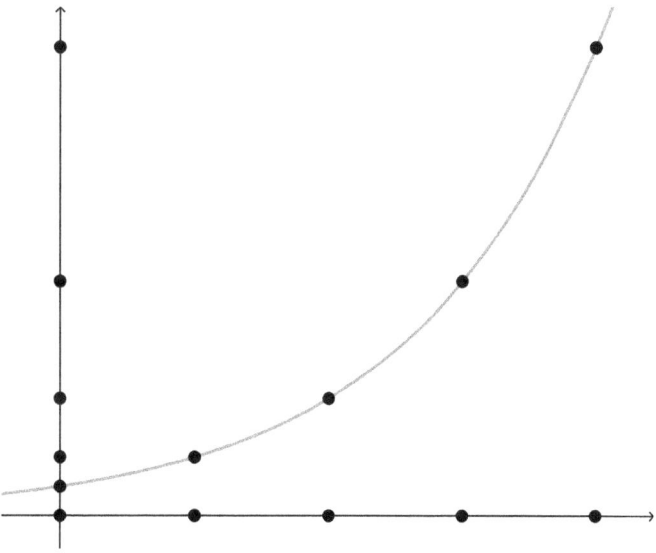

Ex.: Graph the population of a small town over time as its population decreases from 3408 by 2.5% per year.

| x | | | | | |
|---|---|---|---|---|---|
| y | | | | | |

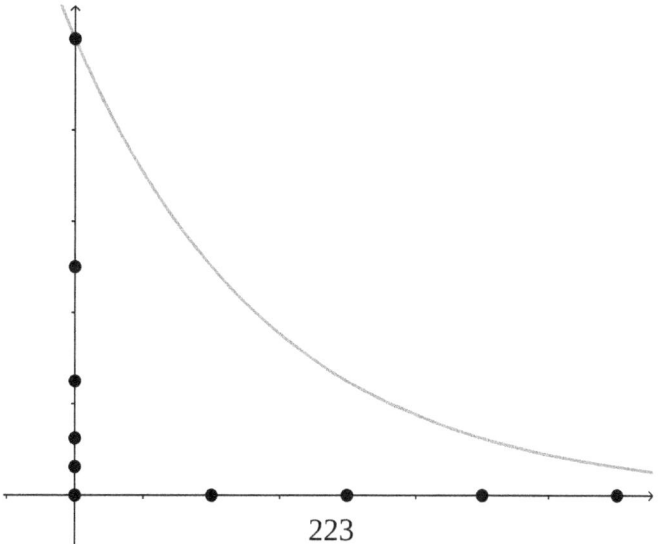

1    The formula for accumulated interest, $A = P\left(1 + r/n\right)^{nt}$ and formula for continuous compounded interest, $A = Pe^{rt}$, are examples what type of equation?

2    All quantities that grow at a constant growth rate also have a what other essential characteristic?

3    The formula for doubling time only depends on what other variable?

4    The formula for doubling time is $T \approx \dfrac{0.69}{r}$ since 0.69 is approximately what quantity?

5    As pedestrian deaths from automobile drivers grows at 10% per year, how long until the number of deaths doubles?

6    As the profits of a major corporation grows at 23% per year, how long until the profits double?

**7**   Graph the population of a small town over time as its population increases from 1300 by 3% per year.

| x | | | | | |
|---|---|---|---|---|---|
| y | | | | | |

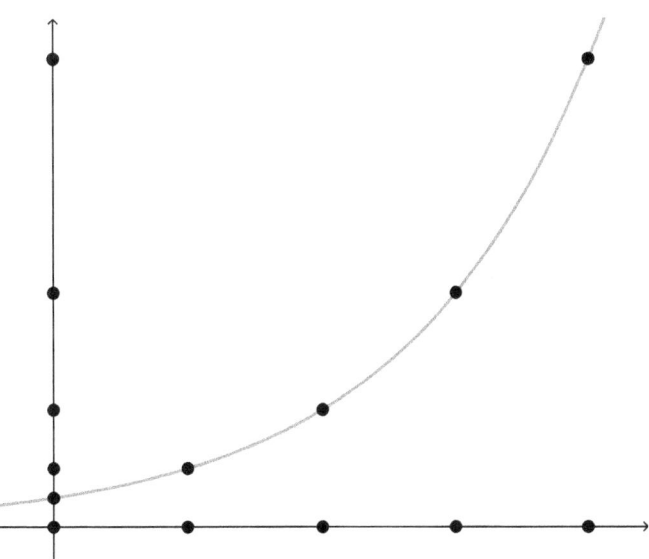

**8**   Graph the population of a small town over time as its population decreases from 2768 by 2.5% per year.

| x | | | | | |
|---|---|---|---|---|---|
| y | | | | | |

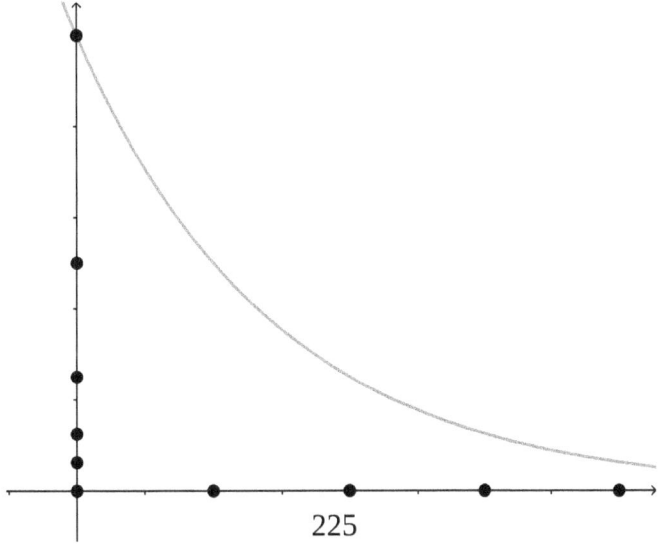

225

1    The formula for accumulated interest, $A = P(1 + r/n)^{nt}$ and formula for continuous compounded interest, $A = Pe^{rt}$, are examples what type of equation?

2    All quantities that grow at a constant growth rate also have a what other essential characteristic?

3    The formula for doubling time only depends on what other value?

4    The formula for doubling time is $T \approx \dfrac{0.69}{r}$ since 0.69 is approximately what quantity?

5    As cyclist deaths from automobile drivers grows at 13% per year, how long until the number of deaths doubles?

6    As the profits of a credit card corporation grows at 19% per year, how long until the profits double?

7    Graph the population of a small town over time as its population increases from 260 by 5% per year.

| x | | | | | |
|---|---|---|---|---|---|
| y | | | | | |

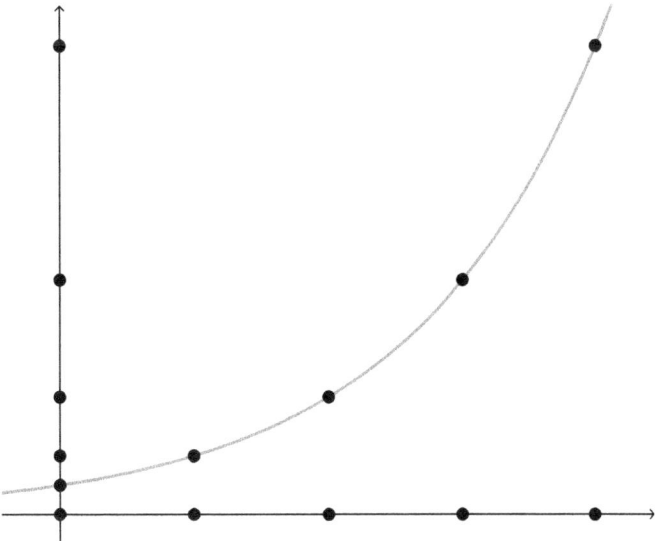

**8**     Graph the population of a small town over time as its population decreases from 3472 by 7% per year.

| x | | | | | |
|---|---|---|---|---|---|
| y | | | | | |

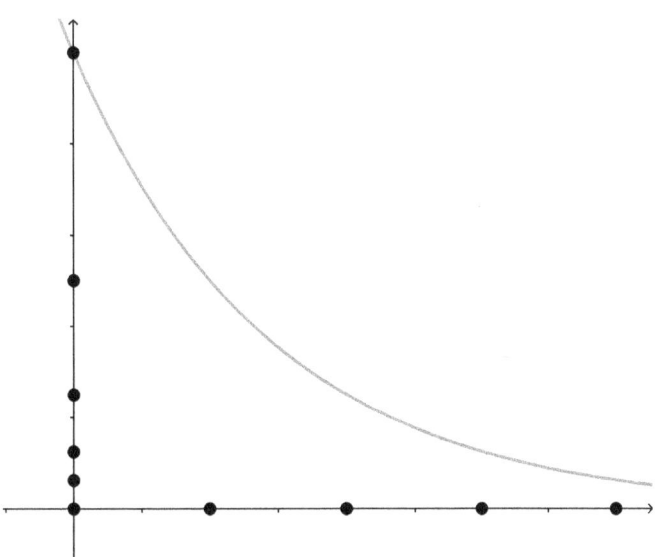

1    Derive the formula for the doubling time given the growth rate, $r$. Use the formula for continuous exponential growth: $A = Pe^{rt}$. (Hint: Since the doubling time is constant, for any beginning value, $P$, the ending value will be $A = 2P$.)

2    How long will it take a state growing at 5% per year to grow from 8 million people to 12 million people? (Hint: Use the formula for continuous exponential growth: $A = Pe^{rt}$.)

1   As credit card debt grows at 18% per year, how long until the debt doubles?

2   How long will it take for a credit card debt being charged 8% interest to grow from $7,500 to $17,250? (Hint: Use the formula for continuous exponential growth: $A = P e^{rt}$.)

3   Solve the formula for continuous exponential growth, $A = P e^{rt}$, for $t$.